GENERAL MOTORS

ADVANCED GENERATION
DIESEL-ELECTRIC AND ELECTRIC LOCOMOTIVES
– THE SECOND AND THIRD GENERATION LOCOMOTIVES –

JAMES W. KERR

A
DELTA
PUBLICATION

83374-B6

GENERAL MOTORS

ADVANCED GENERATION DIESEL-ELECTRIC AND ELECTRIC LOCOMOTIVE

– THE SECOND AND THIRD GENERATION LOCOMOTIVES –

Special thanks are extended to all those who have co-operated and assisted to help make this publication a reality.

ISBN-0-919295-18-5

COPYRIGHT

 Published by Delta Publications Associates Division of DPA-LTA Enterprises Inc.
Printed in Canada

AUTHOR, PRESIDENT AND CHIEF EXECUTIVE OFFICER
JAMES W. KERR

PUBLISHER

DELTA PUBLICATIONS ASSOCIATES DIVISION
DPA-LTA ENTERPRISES INC.

P.O. Box 377
Alburg, VT 05440

P.O. Box 100 - Station "R"
MONTREAL, Quebec Canada
H2S 3K6

CONTENTS

GENERAL INTRODUCTION TO GM CURRENT LOCOMOTIVES AND THEIR HIGHLIGHTS

General Motors Corporation, the World's largest industrial enterprise, with annual sales greatly exceeding one hundred billion dollars, has enjoyed for several decades the distinction of being the largest diesel-electric locomotive builder on Earth. GM has two locomotive production facilities, the larger plant, at La Grange, Illinois, has since the 1930's, produced the greater share of over 50,000 diesel-electric locomotives operating in 57 countries, and is known as the Electro-Motive Division (EMD). The smaller production facility somewhat newer, built in 1950, is known as the Diesel Division of General Motors of Canada Limited, at London, Ontario, (DD-GMCL), but since 1979 has supplied over 50% of all GM locomotives in the World for the export market. DD-GMCL is proud of the enviable claim that since 1961, every locomotive is still in service, except for a few wrecked beyond economical repair. What greater tribute could be paid to DD-GMCL than the acquired reputation of the top quality locomotive production plant in the World? The combined total locomotives built by these two plants is about three times the global production of the closest competitor. Not ever resting on it's well deserved reputation, GM continually maintains an on-going relentless research, development and testing programme for it's locomotives; built for all weather conditions and terrain in the World, preserving it's reputation and market penetration.

The first GM diesel-electric locomotives built in the 1930's, led to the development and perfection of the 567 Series engines, the First Generation diesels, which in 1965 evolved into the 645 Series engines, the Second Generation diesels. Advanced improvements have just recently been incorporated into yet a new 710 Series. Third Generation engine, exemplified in the stellar SD60 Model. Unmatched in the World for unprecedented efficiency and operation, the SD60 Model, a heavy duty 3800hp diesel-electric locomotive incorporates features as SuperSeries traction, and ceramic micro-processor on-board capable of logic functions. As if this were not enough, GM's new 710G3 Series diesel can squeeze the maximum energy out of a drop of diesel oil, and research tends to confirm that in the future, several hundred additional horsepower could become available. The efficency may be compared this way : Three SD60 3800hp Models can outperform four SD40 3000hp units. This is a product of approximately a billion dollars invested in continual research and development.

In North America, GM diesel-electric locomotives have been fine-tuned to perfection, since oil energy is sufficiently less expensive than electricity, compared to Europe, where the opposite in locomotive development occured. Always prepared for a challenge in the event electric locomotives came to favor, GM negotiated an agreement to manufacture technology from one of the World's leading electric locomotive builders, ASEA of Sweden. A quantity of such electric locomotives have been built in both plants, for operation by Amtrak, and the British Columbia Railway. Of notable interest, in the 1960's, GM built it's first electric locomotive, in Canada, for A-T-O operation at an iron ore mine in Northern Quebec.

James W. Kerr

EMD's Super Series Adhesion Control System Boosts Rail Adhesion 33%.

What is Super Series?

GM's Super Series is a dramatic new advance in wheel slip detection and control. Using the latest radar and solid-state integrated circuit technology, Super Series raises the attainable adhesion level by 33%.

To perform effectively, today's high horsepower locomotives require a high adhesion capability. Without higher adhesion, the benefit of high horsepower is lost at low speeds. Super Series provides the higher adhesion capability to permit operation at the minimum continuous speed/tractive effort level without increasing the locomotive weight per axle.

How Does it Work?

Super Series adapts a principle well known to automobile drivers in winter. An experienced driver knows that even in deep snow, he has to have a small amount of wheel slip to keep the car moving. If he lets up on the accelerator, the car will bog down. If he has too much wheel slip, the car will stop or slide sideways. In sum, a certain amount of wheel slip is a good thing. None at all, or too much, is not.

Traditional wheel slip systems are correcting-type systems that detect a wheel slip and then reduce the locomotive power to stop the slip. By contrast, the Super Series controlled wheel slip system allows the wheel to creep by controlling the locomotive power.

The key to the system is accurate measurement of the true ground speed of the locomotive. To establish a true train speed reference signal, the system uses a groundspeed radar similar to that used by the police.

A small radar device bounces a signal off the roadbed. Based on the shift in frequency from the signals transmitted and received (Doppler effect), the true ground speed is determined. The electronic control modules control the maximum allowable wheel creep in the system.

What are the Benefits?

Super Series provides 24% continuous adhesion even on poor rail conditions. It gives the engineer smooth-as-silk control to prevent train handling problems due to improper control of the load.

With controlled creep, Super Series provides high adhesion even without sand. The system reduces sand usage an average of 70%. The system produces less wheel wear than on comparable high-horsepower locomotives or on lower horsepower locomotives with older correcting-type wheel slip systems. This also means less ballast cleaning is required.

AR11 Generator

GM's generators set the standard of the industry for road locomotives. Year after dependable year, these sturdy electrical machines provide the power source for traction motor operation.

The AR11 generator is rated at 7020 amperes continuous compared to the 4200 ampere rating of the AR10. This higher rating, in an overall size only slightly larger than the AR10, is accomplished by the use of generator transition. An external contactor connects the output of the two rectifier banks, either in parallel—for the higher current levels needed at low locomotive speeds—or in series, for the higher voltage levels needed at higher speeds. The generator transition feature also provides higher generator efficiency over a broad operating range.

Another design feature that heightens efficiency is thinner insulation which withstands the same voltage stress as the previous thicker insulation. This allows the use of more copper in the stator and additional turns on the rotor poles, reducing excitation current and power requirements.

Cooling has been improved by the thinner insulation and by the addition of fan blades on the rotor assembly.

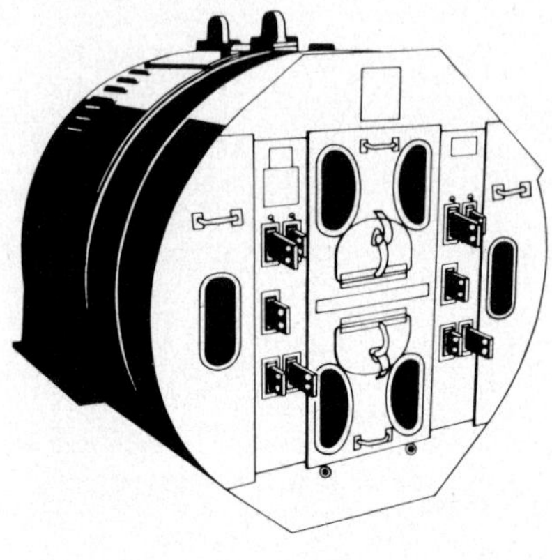

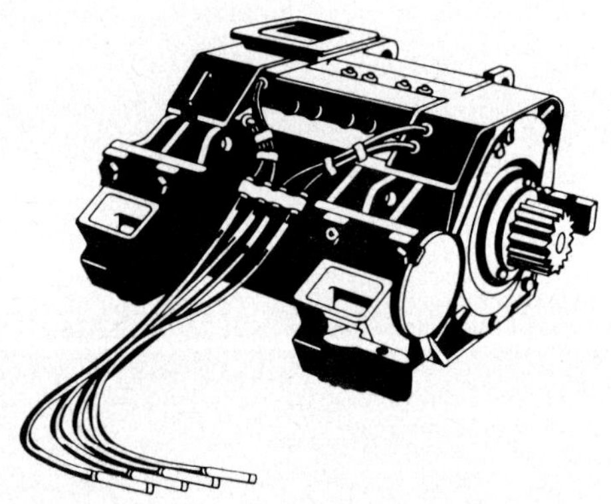

D87 Traction Motor

GM traction motors provide high performance coupled with improved thermal performance at low speeds. The continuous current rating of the latest generation, the D87, has been increased more than 11% compared to the previous generation. This increase was achieved with no increase in overall size.

The D87 features several significant improvements. The armature is built with transposed conductors which reduce eddy currents in the armature coils and heat generated in the motor. This allows the motor to transmit more horsepower. In addition, the copper area in the main field and interpole coils has been increased. The surface area of the main field and interpole coils has been increased to facilitate cooling. Locomotive fuel efficiency has also been improved with a lower horsepower blower and redesigned blower duct.

The mechanical design of the D87 has been improved to handle higher torque and horsepower. The pinion and axle gear have a finer pitch, higher-pressure angle gear system with harder gear tooth surfaces. The larger pinion end bearing has a special crowned inner race for increased load capacity.

Series 645 Diesel Engine

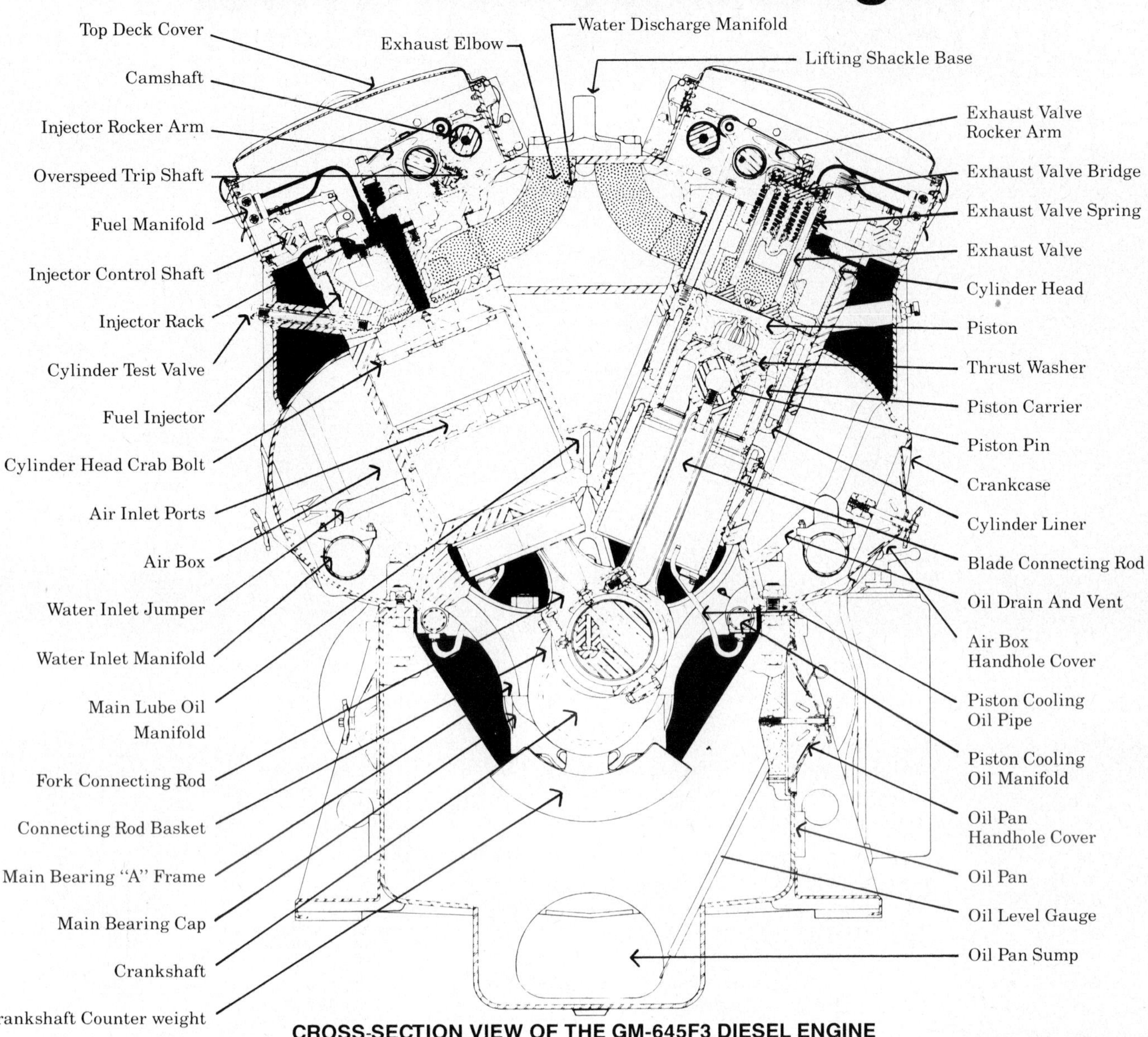

CROSS-SECTION VIEW OF THE GM-645F3 DIESEL ENGINE

Specifications Common to All Series 645 Engines

The GM 645 series diesel engine is a two-stroke cycle, uniflow scavenged, open combustion chamber, poppet valve, 45° Vee design. This 645 engine series is produced in 8, 12 and 16 cylinders, normally aspirated (roots blower scavenged), and 8, 12, 16, and 20 cylinder turbocharged models.

Type	2-cycle—45° Vee
Crankcase and oil pan construction	Welded steel
Bore x stroke	230.2 x 254 mm
Displacement per cyl.	10.58 litres
Operating speed range	310-900 rpm
Piston speed ft./min.	900 rpm—1500
Cylinder air inlet	Ports in cylinder liner
Exhaust	Four valves in cylinder head
Piston cooling	Oil-Direct pressure stream
Main bearing lubrication	Full pressure
Lube oil pumps	Scavenging/pressure/piston cooling/engine driven/positive displacement
Governor	Woodward
Fuel supply pump	Positive displacement/engine driven/or electric motor driven
Fuel injectors	GM unit injectors/needle valve
Engine starting	Air motor/or electric motor
Water pumps	Engine driven/centrifugal
Crankshaft main bearing diameter	19.05 cm
Crankpin diameter	16.51 cm
Piston pin diameter	9.35 cm
Brake mean effective pressure	6.61 kg/cm^2 (normally aspirated model)
Compression Rates — 16:1 — RB	9.91 kg/cm^2 (turbocharged model)
14:5:1 — Turbo	9.35 kg/cm^2 (20 cylinder turbocharged model.)

SERIES 645 TURBOCHARGED DIESEL ENGINE

1. Exhaust Outlet
2. Turbocharger
3. Lube Oil Separator
4. Turbocharger Aftercooler
5. Top Deck Cover
6. Cylinder Relief Valve
7. Crankcase Handhole Cover
8. Oil Pan Handhole Cover
9. Ring Gear
10. Cylinder Head Retainer Plate
11. Rocker Arms - Injector and Exhaust Valves
12. Exhaust Valve Bridge
13. Cooling Water Discharge Elbow
14. Cylinder Head
15. Cooling Water Inlet Jumper Line
16. Cooling Water Manifold
17. Connecting Rod - Blade
18. Connecting Rod - Fork
19. Fork Rod Basket
20. Oil Level Gauge
21. Crankshaft
22. Camshaft
23. Water Discharge Manifold
24. Airbox - Inner Vee
25. Main Lube Oil Gallery
26. "Fire-Ring" Piston
27. Exhaust Valves
28. Cylinder Head Retainer Bolt
29. Fuel Manifold
30. Fuel Injector
31. Piston Carrier
32. Thrust Washer - Carrier/Piston
33. Piston Carrier Insert Bearing
34. "Rocking" Piston Pin
35. Scavenging Oil Pump Intake
36. Oil Pan Sump
37. Main Bearing Cap
38. Connecting Rod Bearing
39. Piston Cooling Oil Manifold
40. Piston Co
41. Main Bea
42. Cranksha
43. Cylinder L
44. Injector C
45. Detectors Low Wate
46. Low Wate
47. Airbox Dr
48. Scavengir

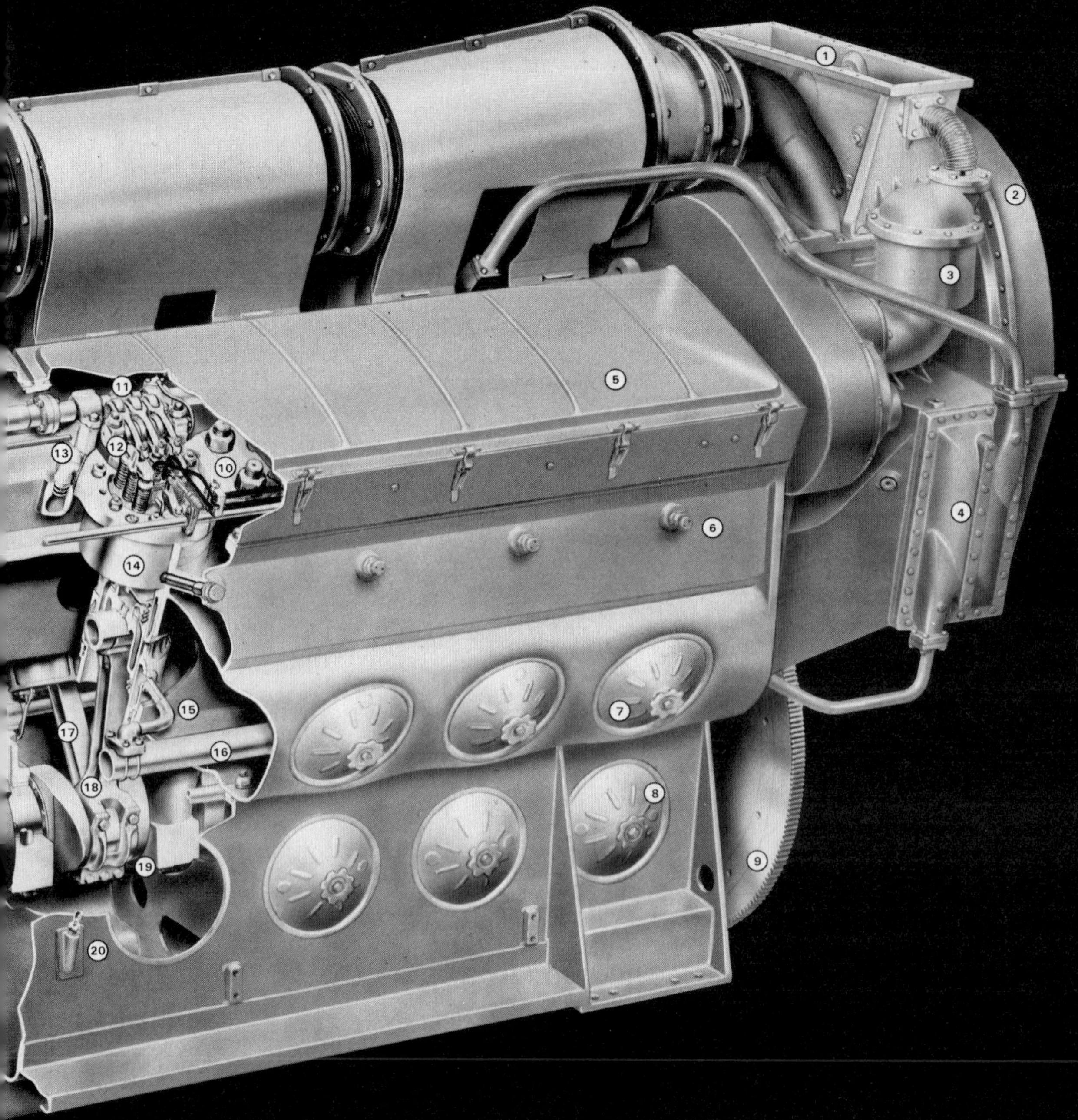

pe

veight

ft Lever

kcase Pressure
Test Cock

p

49. Accessory Drive Coupling
50. Lube Oil Strainer
51. Main Lube/Piston Cooling Oil Pump
52. Cooling Water Pump
53. Governor Terminal Shaft Scale
54. Governor
55. Overspeed Trip Lever
56. Exhaust Manifold
57. Cylinder Exhaust Outlet
58. Exhaust Manifold Heat Shield

710G3 SERIES DIESEL MOTOR

A NEW ENGINE, A NEW ERA

EMD's 60 Series locomotives feature a new diesel engine, the 710G. This engine represents the culmination of an extensive, four-year. $60 million development effort and a $78 million tooling investment.

The new 710G engine is the product of thousands of hours of painstaking research and development by GM. And, as the newest evolution in diesel engine design, it offers the potential for even higher horsepower in the future.

Standars on every 60 Series locomotive, the 710G is a giant stride forward in increased reliability, fuel efficiency and power for diesel engines. The keys to the 710G's impressive performance capabilities are its greater displacement and advanced turbocharger.

By way of comparison to its immediate prodecessor, the 645F, the new 710G features an increased piston stoke, up from the 645F's 10 inches to 11 inches. This increases displacement by 10 percent — from 645 to 710 cubic inches per cylinder.

The 16-cylinder, 710G is rated conservatively at 3950 brake horsepower at 900 rpm (153.0 psi BMEP) and has a compression ratio of 16:1.

But merely introducing more and more powerful engines into the marketplace does not reflect the true nature of EMD's ingenuity and innovativeness. Rather, it is strict commitment to developing products of ever-increasing quality and economy that has made an international leader in the supply of Diesel-electric locomotives to the world.

TURBOCHARGED EFFICIENCY

Even though the 60 Series locomotive is conspicuously more powerful than any of its ancestors, it is also the most economical to operate. A major contributor to this exciting achievement is the 710G's advanced Model G turbocharger.

Through a more streamlined turbocharger design, the flow of gases into the turbine has been made smoother and less restrictive. An improved exhaust diffuser also reduces flow constriction. This improved exhaust flowpath, in combination with a new, high-efficiency compressor makes the Model G turbocharger one of the most efficient in the world.

The result of this progressive engineering is that the Model G turbocharger provides a 15 percent addition over the older model in air flow, thus reducing thermal loading of critical engine components.

This greater air flow, in combination with an increased injection rate from a new 9/16 inch plunger injector, ultimately results in higher air/fuel ratios and significantly improved fuel efficiency.

Not being content with introducing the newest dimension in diesel engines or pioneering the 60 Series microprocessor locomotive control system, the engineers at GM have turned their considerable talents to the support systems of the 60 Series. The resulting improvements have helped reduce parasitic losses in the new locomotives to less than 3 percent of engine ratings. They include:

- increased radiator capacity, combined with other engine efficiency improvements, to result in reduced demands on the cooling system. This in turn reduces cooling fan horsepower requirements;
- two-speed AC cooling fans to provide five cooling air flow rates;
- cooling fans which cycle to conserve energy;
- improved air ducting to the traction motor blowers, requiring less blower horsepower;
- the control of full air flow to the motor, providing it only when required;
- a control which runs the air compressor only when air is needed; and,
- during dynamic braking, a control which idles the engine at the lowest possible speed for motor cooling and fuel conservation.

In addition, many features, including increased capacity lube oil and fuel filters, have been designed into the 60 Series to increase the service period to the next required maintenance.

A new traction motor, the D87A, emphasizes an increased brush face area, improved motor ventilation and a new automatically applied polyimide film insulation system. These advances permit higher horsepower without increasing motor size. Standard on 60 Series locomotives, the D87A is also retrofittable on older models.

By combining these support system improvements with the new 710G diesel engine and the increased control abilities of the onboard microprocessors, the GM 60 Series locomotives become capable of reaching new levels of fuel economy and power while requiring less maintenance than any other locomotive in the history of the industry.

FOUR MODELS

The 60 Series consists of four mainline models, patterned after the SD50 and GP50, which were introduced in 1980. They are:

- The SD60 — a 390,000 lb. (176,900 kg.) 6-axle locomotive intended for heavy-duty drag operation or medium speed freight trains in domestic type mainline service;
- The SD59 — a 6 axle, lower horsepower locomotive for intermediate service;
- The GP60 — a 260,000 lb. (117,935 kg.) 4-axle locomotive for intermediate and high speed service; and,
- The GP59 — a 4 axle, lower horsepower locomotive for intermediate service.

GM has done much more than merely keep up with the rapid pace of technological advancement within the railroad industry, setting new standards for superior performance, reliability and economy with the 60 Series locomotives.

A SHOWCASE OF IMPROVEMENTS

The 60 Series is the latest grand scale achievement from GM, the company that has been behind almost every significant advance in the locomotive industry since its inception in 1922. The 60 Series takes advantage of recent GM breakthroughs in diesel and microprocessor technology, and also incorporates many important gains in support systems.

ON TRACK WITH MICROPROCESSORS

The rapidly advancing technology of microprocessors is having a substantial impact upon almost every sector of American work life. The introduction of microprocessors into the 60 Series locomotive is the fruit of several years of solid engineering effort by GM, and places in our customers'hands the most advanced complete control system the railroads have ever witnessed.

To offset the exceptionally computer-hostile environment of a locomotive (temperature extremes, iron and silica dirt), all chips in the control system are ceramic-based. At or near military grade, they are able to pass reliability tests over the entire temperature range from —85°F to 257°F. A specially rugged grade of connector is likewise specified.

When we design a system to outperform its predecessor, we also make it simpler. The new micro control system on the 60 Series locomotives incorporates a sharply reduced number of components. This not only increases the innate reliability of the system, it decreases the size of your necessary spare parts inventory.

MICRO CONTROLS, MASSIVE BENEFITS

The new control systems in the 60 Series locomotives involve three Motorola 6803 microprocessors, each controlling a separate function: logic, excitation, or diagnostics and display.

The logic computer controls such functions as engine speed, locomotive direction and traction motor switching, previously done with conventional relay logic and wire harness interconnections. Inputs come from such locomotive controls as the throttle. The computer in turn operates the power contractors, engine governor speed solenoids, switchgear and other control devices. It also tells the excitation computer what level of traction power or dynamic braking the locomotive engineer has selected.

The excitation system controls such items as the Super Series wheel creep control, dynamic brake and the latest fuel economy features. In the past, this was accomplished by analog and digital integrated circuits on a variety of plug-in modular circuit boards. The excitation computer receives throttle and brake information from the logic system and established power or brake reference limits. Comparison with feedback information from the main generator, traction motors and dynamic brake system determines when and how power is adjusted.

The diagnostic and display system replaces conventional fault annunciators and indicator lights with a four-line display panel. Software allows many new fault and status indicators to be added as well, and the system is user friendly. It displays conditions in English, not computer codes which require a combersome manual to decipher.

This system also monitors locomotive operation continuously, detecting abnormal conditions, recording the condition of the locomotive at the time the fault occurred and, in some cases, initiating corrective action.

The menu-driven diagnostic system can analyse fault conditions and provide maintenance workers with detailed information on where the problem is right on the locomotive. The display can also be used as a digital voltmeter, displaying up to 12 system parameters simultaneously in real time.

In addition, self-test features make it possible to qualify electrical systems, including radar, contactors and the computer itself. These self-tests also improve service reliability by preventing the dispatch of locomotives with system malfunctions.

The 60 Series micro-controls also add to the reliability of the locomotive by avoiding unnecessary engine shut - downs. Older control systems automatically shut the engine down when certain generic types of faults are detected. By contrast, the 60 Series microprocessor can analyze the fault more precisely and decide whether the problem can be dealt with in some other way.

For example, if the cooling water overheats, power on the 60 Series is reduced to keep temperatures within acceptable limits. This contrasts with the equipment of other locomotive manufacturers that cuts power any time a cooling fan malfunctions — whether or not the engine actually overheats.

THE ARCHIVE MODULE: WHAT, WHERE AND HOW

The diagnostic and display network of the 60 Series microprocessor controls includes the archive module, a system that continuously monitors and records locomotive operational data, such as mileage, kilowatt-hours and duty cycle in real time. Its chief purpose is to augment the diagnostic system's detection of an abnormal running condition by permanently recording the locomotive's status at the time that the fault occurred. This valuable, and previously unavailable, information will help speed repair time and reduce costs after a failure.

Another advantage of the archive module's record of locomotive operation is that it enables routine maintenance and overhauls to be scheduled when they are actually needed rather than on fixed time or mileage intervals.

Up until this development, maintenance schedules have reflected the average requirements of the locomotive fleet. But the actual duty cycles of individual locomotives often vary widely from the average, so the majority of the fleet receives maintenance either sooner than necessary or not soon enough.

With the microprocessor-based archive module, the maintenance schedule can be based on actual work performed, as reflected in horsepower-hours or kilowatt-hours. This maximizes the life of components requiring periodic replacement while safeguarding against continued operation with worn out equipment.

The continuous duty record compiled by the archive module provides another vital advantage for the 60 Series locomotives: analysis of this data can tell the railroad operator whether a given train has been over or underpowered. Further inspection makes it possible to adjust train makeup and size to provide the most efficient train operation and locomotive utilization over various routes.

Permanence of the archive module's record is assured by battery back-up, which allows the module to be removed from the chassis without risking the loss of stored information.

PHOTOGRAPH SECTION

Diesel-Electric and Electric Locomotives

Traction Motor Cabless Booster Unit For a Switcher Locomotive (Zéro Horsepower).

 SW1000 Model 1000 hp Diesel-Electric Switcher.

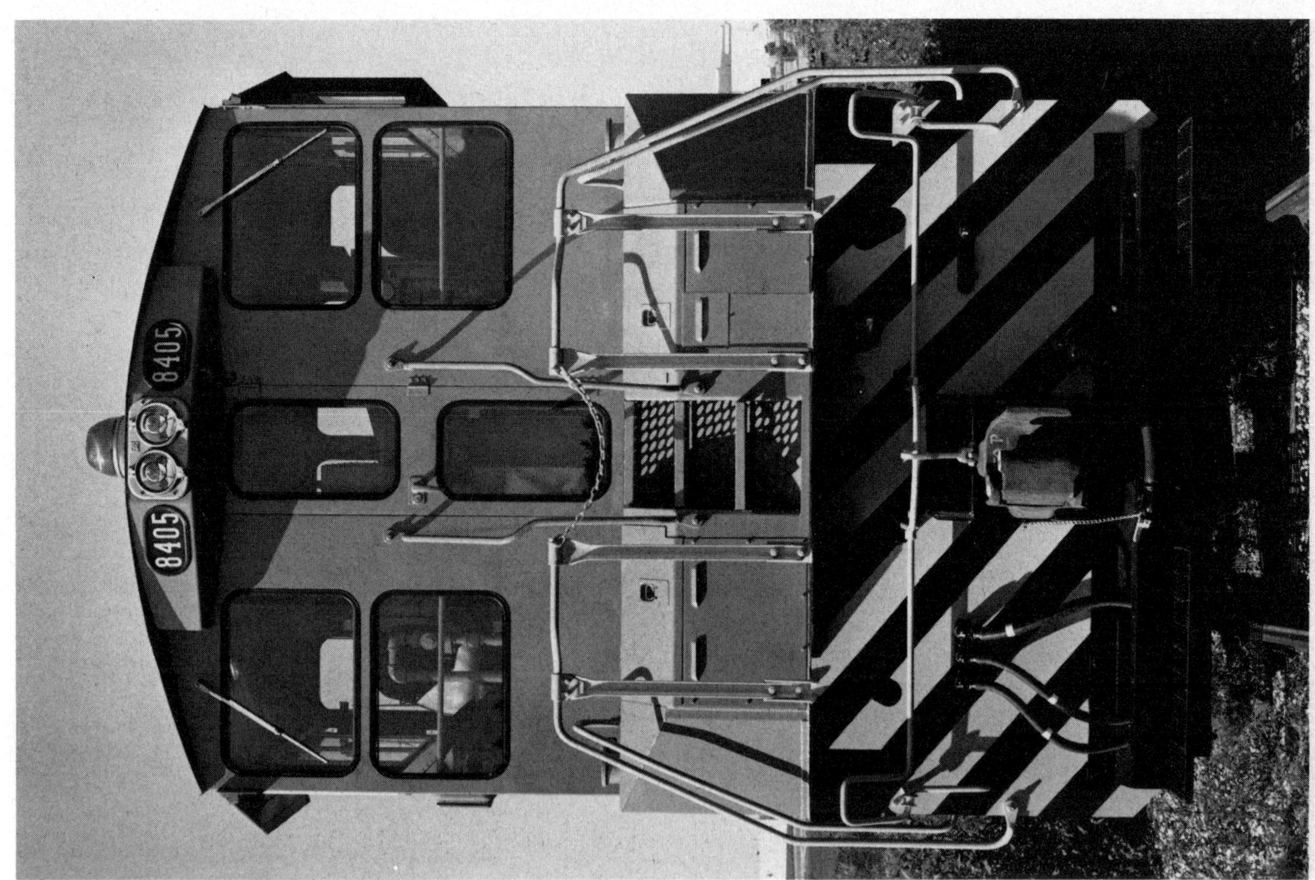

MP15AC Model 1500 hp Diesel-Electric Multi-Purpose Locomotive (Front-View).

SW1001 Model 1000 hp Diesel-Electric Switcher.

MP15AC Model 1500 hp Diesel-Electric Multi-Purpose Locomotive.

 MP15AC Model 1500 hp Diesel-Electric Multi-Purpose Locomotive.

GP15-1 Model 1500 hp Diesel-Electric General Purpose Locomotive.

SD38-2 Model 2000 hp Diesel-Electric Special Duty Locomotive.

SD38-2 Model 2000 hp Diesel-Electric Special Duty Locomotive.

SD38-2 Model 2000 hp Diesel-Electric Special Duty Locomotive (Front View).

 SD38-2 Model 2000 hp Diesel-Electric Special Duty Locomotive.

GP38 Model 2000 hp Diesel-Electric General Purpose Locomotive.

GP38 Model 2000 hp Diesel-Electric General Purpose Locomotives.

GP38 Model 2000 hp Diesel-Electric General Purpose Locomotive.

GP38 Model 2000 hp Diesel-Electric General Purpose Locomotive.

GP38-2 Model 2000 hp Diesel-Electric General Purpose Locomotive.

GP38-2 Model 2000 hp Diesel-Electric General Purpose Locomotive.

GP38-2 DCC Model 2000 hp Diesel-Electric General Purpose Locomotive Deluxe Canadian Cab.

GP39-2 Model 2300 hp Diesel-Electric General Purpose Locomotive SFSPC

SD39 Model 2300 hp Diesel-Electric Special Duty Locomotive. SFSPC

GP40-2 DCC Model 3000 hp Diesel-Electric General Purpose Locomotive - Deluxe Canadian Cab.

GP40 Model 3000 hp Diesel-Electric General Purpose Locomotive.

 GP40TC Model 3000 hp Diesel-Electric General Purpose Passenger Locomotive.

GP40P-2 Model 3000 hp Diesel-Electric General Purpose Passenger Locomotives. SFSPC

GP40-2 DCC Model 3000 hp Diesel-Electric General Purpose Locomotive - Deluxe Canadian Cab.

F40PH-2 Model 3000 hp Diesel-Electric Passenger Locomotive. (Front View).

GP40-2 DCC Model 3000 hp Diesel-Electric General Purpose Locomotive Deluxe Canadian Cab (Front View).

F40PH Model 3000 hp Diesel-Electric Streamlined Passenger Locomotive.

F40PH Model 3000 hp Diesel-Electric Streamlined Passenger Locomotive.

SD40 Model 3000 hp Diesel-Electric Special Duty Locomotive.

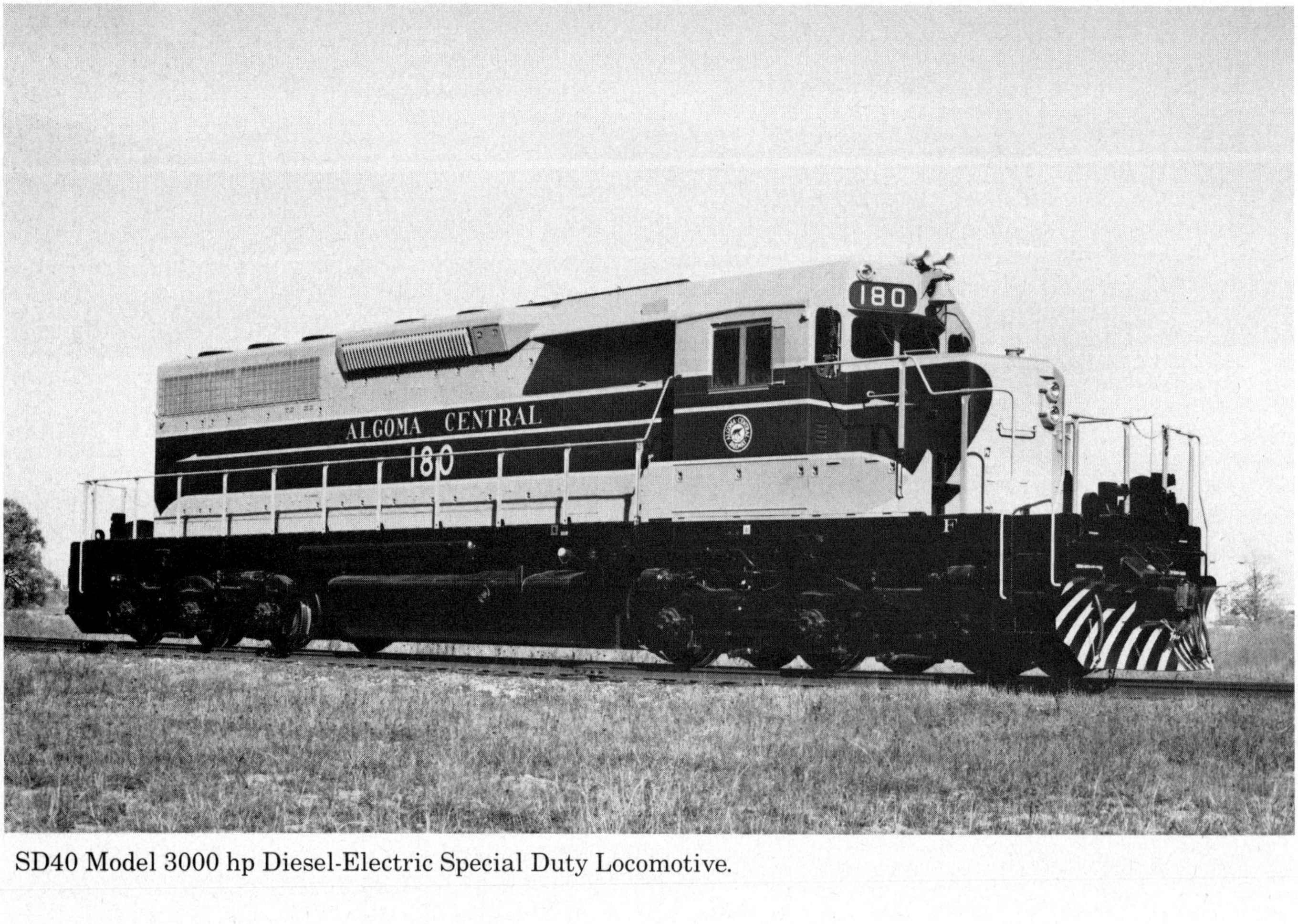

 SD40 Model 3000 hp Diesel-Electric Special Duty Locomotive.

SD40 Model 3000 hp Diesel-Electric Special Duty Locomotive.

SD40 Model 3000 hp Diesel-Electric Special Duty Locomotive.

SD40 Model 3000 hp Diesel-Electric Special Duty Locomotive.

 SD40 Model 3000 hp Diesel-Electric Special Duty Locomotive.

SD40 Model 3000 hp Diesel-Electric Special Duty Locomotive.

SD40 Model 3000 hp Diesel-Electric Special Duty Locomotive.

SDP40 Model 3000 hp Diesel-Electric Special Duty Passenger Locomotive.

 SD40T-2 Model 3000 hp Diesel-Electric Special Duty Locomotive with Lower Air Intake.

SD40-2 Model 3000 hp Diesel-Electric Special Duty Locomotive.

SD40-2 Model 3000 hp Diesel-Electric Special Duty Locomotive.

SD40-2 Model 3000 hp Diesel-Electric Special Duty Locomotive.

 SD40-2 Model 3000 hp Diesel-Electric Special Duty Locomotive.

SD40-2 Model 3000 hp Diesel-Electric Special Duty Locomotive.

SD40-2 Model 3000 hp Diesel-Electric Special Duty Locomotive.

SD40 Model 3000 hp Diesel-Electric Special Duty Locomotive with the High Short Hood.

SD40-2 Model 3000 hp Diesel-Electric Special Duty Locomotive with the High Short Hood.
Long Hood Front

SD40-2 Model 3000 hp Diesel-Electric Special Duty Locomotive.

SD40-2 Model 3000 hp Diesel-Electric Special Duty Locomotive.

SD40-2 Model 3000 hp Diesel-Electric Special Duty Locomotive Built by DD-GMCL.

 SD40-2 Model 3000 hp Diesel-Electric Special Duty Locomotive Built by DD-GMCL.

SD40-2 Model 3000 hp Diesel-Electric Special Duty Locomotive Built by DD-GMCL.

SD40-2 Model 3000 hp Diesel-Electric Special Duty Locomotive Built by DD-GMCL.

SD40-2 Model 3000 hp Diesel-Electric Special Duty Locomotive Built by DD-GMCL (Front View).

SD40-2 Model 3000 hp Diesel-Electric Special Duty Locomotive Built by DD-GMCL (Rear View).

SD40-2 Model 3000 hp Diesel-Electric Special Duty Locomotive.

SD40-2 Model 3000 hp Diesel-Electric Special Duty Locomotive.

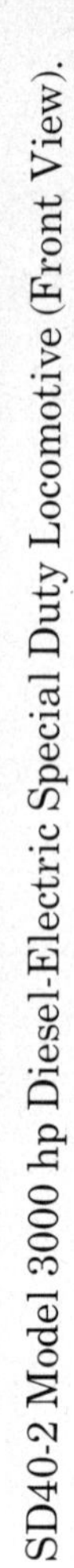
SD40-2 Model 3000 hp Diesel-Electric Special Duty Locomotive (Front View).

SD40-2 Model 3000 hp Diesel-Electric Special Duty Locomotive (Roof View).

SD40-2 Model 3000 hp Diesel-Electric Special Duty Locomotive.

SD40-2 Model 3000 hp Diesel-Electric Special Duty Locomotive.

SD40-2 Model 3000 hp Diesel-Electric Special Duty Locomotive.

SD40-2 Model 3000 hp Diesel-Electric Special Duty Locomotive. Engineer's Console.

SD40-2 DCC Model 3000 hp Diesel-Electric Special Duty Locomotive - Deluxe Canadian Cab.

SD40-2 DCC Model 3000 hp Diesel-Electric Special Duty Locomotive - Deluxe Canadian Cab.

SD40-2 DCC Model 3000 hp Diesel-Electric Special Duty Locomotive Deluxe Canadian Cab Engineer's Console.

SD40-2 DCC Model 3000 hp Diesel-Electric Special Duty Locomotive Deluxe Canadian Cab (Front View).

F40C Model 3200 hp Diesel-Electric Streamlined Passenger Locomotive - Commuter Service.

SD45 Model 3600 hp Diesel-Electric Special Duty Locomotive.

F45 Model 3600 hp Diesel-Electric Streamlined Passenger Locomotive.

F45 Model 3600 hp Diesel-Electric Streamlined Passenger Locomotive.

SD45T-2 Model 3600 hp Diesel-Electric Special Duty Locomotive with Lower Air Intake.

DDA-40X Model 6600 hp Worlds' Most Powerful Diesel-Electric Locomotive (has 2 engines).

GP40X Model 3500 hp Diesel-Electric General Purpose Locomotive. SFSPC

 GP50 Model 3500 hp Diesel-Electric General Purpose Locomotive.

SD50F DCC Model 3500 hp Diesel-Electric Special Duty Locomotive - Deluxe Canadian Cab - Engineer's Console.

SD50F DCC Model 3500 hp Diesel-Electric Special Duty Locomotive - Deluxe Canadian Cab.

SD50F DCC Model 3500 hp Diesel-Electric Special Duty Locomotive - Deluxe Canadian Cab.

SD50 Model 3500 hp Diesel-Electric Special Duty Locomotive.

SD50 Model 3500 hp Diesel-Electric Special Duty Locomotive.

SD50 Model 3500 hp Diesel-Electric Special Duty Locomotive.

SD60 Model 3500 hp Diesel-Electric Special Duty Locomotive.

SD60F DCC Model 3800 hp Diesel-Electric Special Duty Streamlined Freight Locomotive - Deluxe Canadian Cab.

SD60F DCC Model 3800 hp Diesel-Electric Special Duty Streamlined Freight Locomotive - Deluxe Canadian Cab, Engineer's Console.

SW1200 A-T-O Model 1200 hp Electric Locomotive for Automatic - Train - Operation
(The first Electric Locomotive built by GM was built in Canada.)

GF6C Model 6000 hp Electric Locomotive ASEA-GM with the panthograph in a raised position.

GF6C Model 6000 hp Electric Locomotive ASEA-GM.

GF6C Model 6000 hp Electric Locomotive ASEA-GM.

GF6C Model 6000 hp Electric Locomotive (Front-View) ASEA-GM.

 GF6C Model 6000 hp Electric Locomotive ASEA-GM.

GM6C Model 6000 hp Electric Locomotive ASEA-GM-EMD Demonstrator.

AEM7 Model 7000 hp Electric Locomotive for Fast Passenger Train Services. ASEA-GM.

COLOR SECTION

Locomotives identified by Model, Horsepower and Operator (14 Diesels, 2 Electrics).

MP15AC	Model,	1500 hp,	National Harbours Board #8405
SD38-2	Model,	2000 hp,	Northern Alberta Railway #401
F40PH	Model,	3000 hp,	Government of Ontario Transit #514
F40PH-2	Model,	3000 hp,	Via Rail Canada #6412
SD40-2	Model,	3000 hp,	Canadian Pacific Railway #5943
SD40-2*	Model,	3000 hp,	Atchison, Topeka & Santa Fe Railway #5032
SD40X	Model,	3500 hp,	Kansas City Southern Railway #702
SD50*	Model,	3500 hp,	Canadian National Railway #5404
GP59	Model,	3000 hp,	GM – Electro Motive Division #8
GP60	Model,	3800 hp,	GM – Electro Motive Division #5
SD60	Model,	3800 hp,	GM – Electro Motive Division #1
SD60F*	Model,	3800 hp,	Canadian National Railway #9900
SD60	Model,	3800 hp,	Chicago & North Western Transportation #8001
SD60	Model,	3800 hp,	Union Pacific Railroad #6005
GM10B	Electric,	10,000 hp,	ASEA-GM Electric Locomotive #1976
GF6C	Electric,	6,000 hp,	ASEA-GMCL-DD British Columbia Railway Electric Locomotive #6006

*Deluxe Canadian Cab

8405

401
NAR
401
NAR
NORTHERN ALBERTA

514
514
GCE-430e
Government of Ontario Transit

6412
6412
6412
VIA
VIA

5943
CP Rail
5943

5032
5032
5032
SANTA FE
Santa Fe

702
702
702
KCS

5404
5404
GF-636a

EMD 8
EMD 8
Power of Tomorrow, Today
GP59
ELECTRO-MOTIVE

EMD 5
EMD 5
GM
EMD 5
Power of Tomorrow, Today
GP60
ELECTRO-MOTIVE

EMD 1
EMD 1
GM
EMD 1
Power of Tomorrow, Today
ELECTRO-MOTIVE

9900
9900
GF-638a
M

6005
UNION PACIFIC
6005
UNION PACIFIC

1976
1976
1976
GM
ELECTRO-MOTIVE
TEST CAR
ELECTRO-MOTIVE

6006
6006
RAILWAY
KEEP OFF ROOF
HIGH VOLTAGE

ENGINEERING SPECIFICATION DRAWINGS – DIESEL-ELECTRIC LOCOMOTIVES

MODEL SW1001 1000 HP – 115 TON LOCOMOTIVE

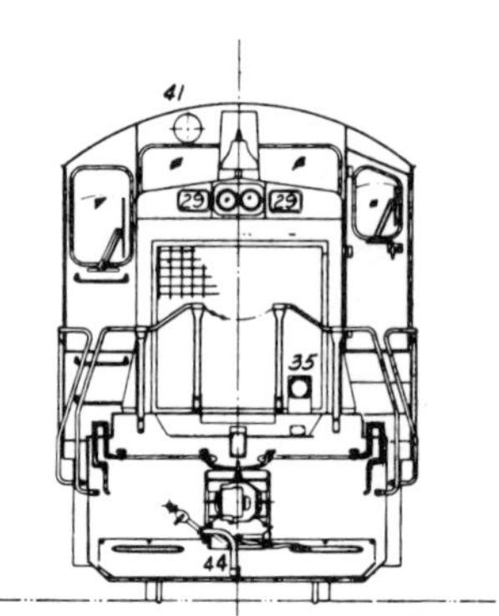

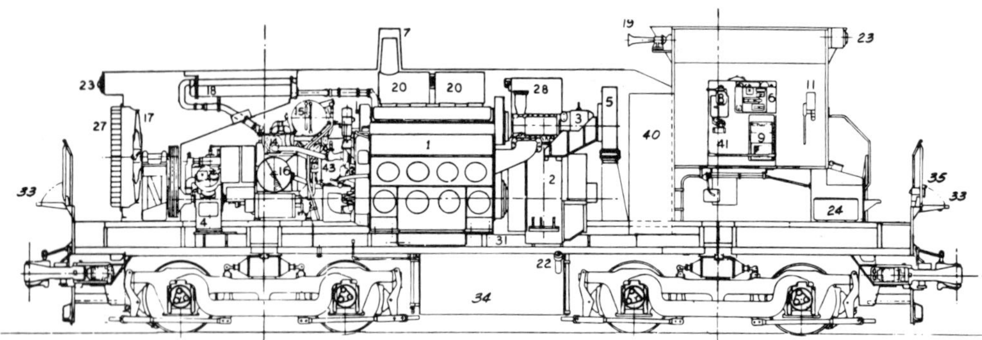

LOCOMOTIVE WITH EXTRAS

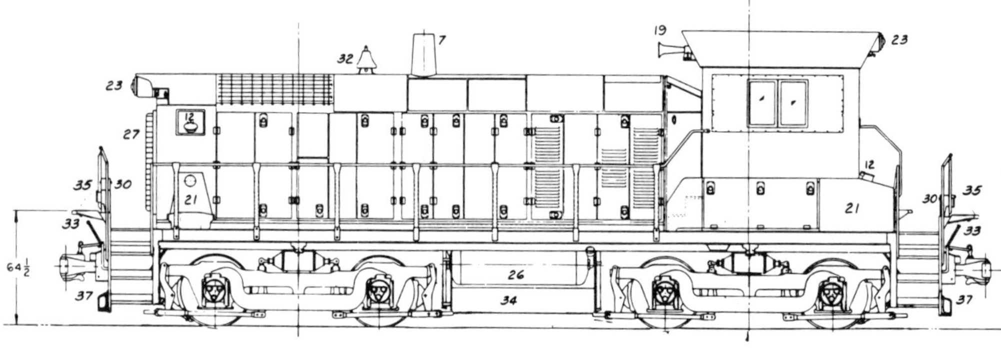

LEGEND

1. ENGINE – EMD MODEL 8-645E
2. MAIN GENERATOR
3. AUXILIARY GENERATOR 10 KW
4. AIR COMPRESSOR
5. TRACTION MOTOR BLOWER
6. ENGINEER'S CONTROL
7. EXHAUST STACK
8. AIR BRAKE VALVE
9. CAB HEATER
10. SLIDING SEAT
11. HAND BRAKE
12. SAND BOX FILLER
13. LUBE OIL FILLER
14. LUBE OIL COOLER
15. ENGINE WATER TANK
16. LOAD REGULATOR
17. FAN
18. RADIATOR
19. HORN
20. EXHAUST MANIFOLD
21. SAND BOX 30 CU. FT. TOTAL
22. FUEL FILLER
23. HEADLIGHT TWIN SEALED BEAM
24. BATTERIES
25. FUEL TANK 600 GALLONS
26. MAIN AIR RESERVOIR
27. AIR INLET & SHUTTERS
28. ENGINE AIR FILTER

*29. NUMBER BOX

*30. RAMP LIGHT

31. TRACTION MOTOR AIR DUCT
32. BELL

*33. M.U. END ARRANGEMENT (GP-TYPE)

*34. FUEL TANK 900 GALLONS

*35. M.U. RECEPTACLE

*36. WATER COOLER

37. PILOT & FOOTBOARDS
38. REF. NOT USED

*39. DUPLEX CONTROLLER

40. ELECTRICAL CABINET
41. CAB VENTILATOR
42. ENGINE CONTROL PANEL
43. ENGINE START SWITCH

*44. M.U. END ARRANGEMENT (SW-TYPE)

45. CONTROL HANDLE ARRANGEMENT (SW1200 TYPE)

*MODIFICATIONS

MODEL MP15T — 1500 HP MULTI-PURPOSE LOCOMOTIVE

LEGEND

1. Engine - 8-645E3C (1500 HP)
2. Main Generator AR10/D18
3. Auxiliary Generator 18 KW-A.C.
4. Traction Motor/Generator Blower
5. Engine Air Filters
6. Air Compressor (WBO Shown)
7. Radiators (6 IN/6 ROW - 4 Total)
8. Cooling Air Shutters
9. Cooling Fan (48 In. Dia. "Q" Type - 8 Blade)
10. Engine Water Tank
11. Batteries (MS 280)
12. Sandbox (Cab End - 15 Cu. Ft.)
13. Sandbox (Hood End - 15 Cu. Ft.)
14. Coupler (E-Type)
15. Draft Gear (MS 485 - 6A)
16. Fuel Tank (1100 Gal.)
17. Main Air Reservoirs
18. Electrical Control Cabinet Access
19. Air Brake Equipment
20. Truck (2 Axle - GP Type - Clasp Brakes Shown)
21. Traction Motor (D77 - Total)
22. Side Footboards & Ladders
23. Handbrake
24. Underframe Mounted Handrails
25. Horn
26. Bell
27. Headlight
28. Engineer's Control Stand
29. Electrical Cabinet Air Filter Box (2 Element)
30. Cab Door
31. Cab Seat
32. Lube Oil Cooler (40 In. Shell & Tube)
33. Engine Exhaust Silencer
34. Number Box
35. Lube Oil Strainer
36. Carbody Impingement Air Filters (15 Total)
37. Lube Oil Filter (5 Total)

*38. MU End Arrangement
*39. MU Receptacle
40. Pilot
*41. Fuel Tank (1600 Gal.)
42. Sand Filler
*43 Class Light
44. Electric Cab Heater
45. Governor
46. Engine Room Partition
47. Cooling Air Inlet Screen
*48. Hopper Compartment
49. Modules
50. Removable Blower Duct
51. Traction Motor Air Duct
52. Side Footboards & Steps
53. Cooling Fan & Radiator Access Door
54. Cooling System Partition
55. Jacking Pads
56. Sand Trap Access Doors
57. A.C. Cabinet
58. 30 In. Primary Fuel Filter
59. Engine Removal Match
60. Traction Motor Blower Enclosure
61. Filtered Air Compartment Access
62. Engine Oil Pan
63. Generator Removal Hatch
64. Cylinder Liner Removal Doors (Hinged)

* Modification

NOTE:
Locomotive Height Tolerance = ± 1-1/2"
Locomotive Width Tolerance = ± 1/2"
Truck Lateral
(Bolsters + Journals w/new trucks = 2-3/4IN) (Nom.)

Locomotive is shown including half variable supplies and in new condition standing still on level and tangent track.

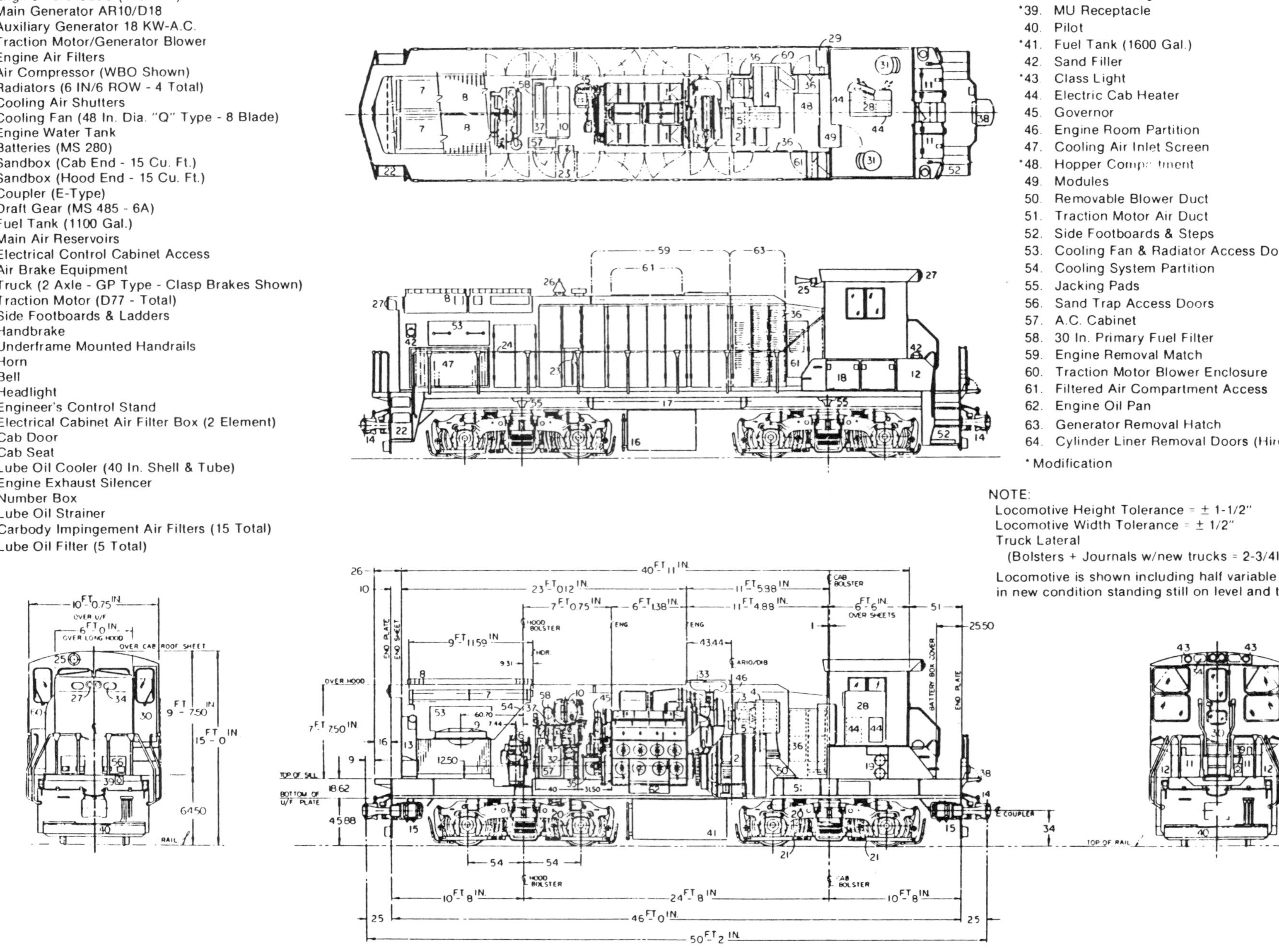

MODEL GP15T — 1500 HP GENERAL PURPOSE LOCOMOTIVE

LEGEND

1. Engine (8-645E3C)
2. Generator/Alternator (AR10/D14)
3. Auxiliary Generator (10 KW)
4. Traction Motor Blower
5. Generator Blower
6. Electrical Control Cabinet
7. Electrical Control Cabinet Air Filter Box
8. Air Compressor (*WBG shown w/special filter)
9. Exhaust Silencer
10. Engine Air Filters (36 in. paper - 6 total)
11. Engine Water Tank
12. Engine Lube Oil Filter (5 element)
13. Engine Lube Oil Cooler (40½ in. shell & tube)
14. 30 in. Primary Fuel Filter
15. Cooling Fan (48 in. "Q" type)
16. Radiators (4 double length - 6 in./6 row)
17. Roof Shutters
18. Cooling Air Inlet Grille
19. Sandbox (28 cu. ft. capacity)
*20. Dynamic Brake
21. Main Generator & Auxiliary Generator Removal Hatch
*22. New Batteries (MS420)
23. Main Air Reservoir
24. Side Footboards & Ladder
25. Coupler ("E" type)
26. Draft Gear (NC391)
27. Air Brake Equipment (26NL)
28. Traction Motor Air Duct
29. Truck (4 wheel GP type-clasp brakes shown)
30. Carbody Air Filters (impingement type - 17 Req'd.)
31. Handbrake Alcove
32. Sandbox Filler
33. Headlight
34. Number Box
35. Horn
36. Fuel Tank (2400 U.S. gallons shown)
37. Brake Cylinder
38. Traction Motors (D77-4 total)
39. Engine Room Partition
40. Air Brake Equipment Access
41. Traction Motor Blower Enclosure
42. Cooling Fan & Radiator Access
43. Filtered Air Compartment Access
44. Battery Access
45. Carbody Filter Access
46. Generator Mounting Access
47. Engine Oil Pan
48. Signal Light
49. Pilot & Hose Storage Rack
50. Cab Heater (hot water)

*Modification

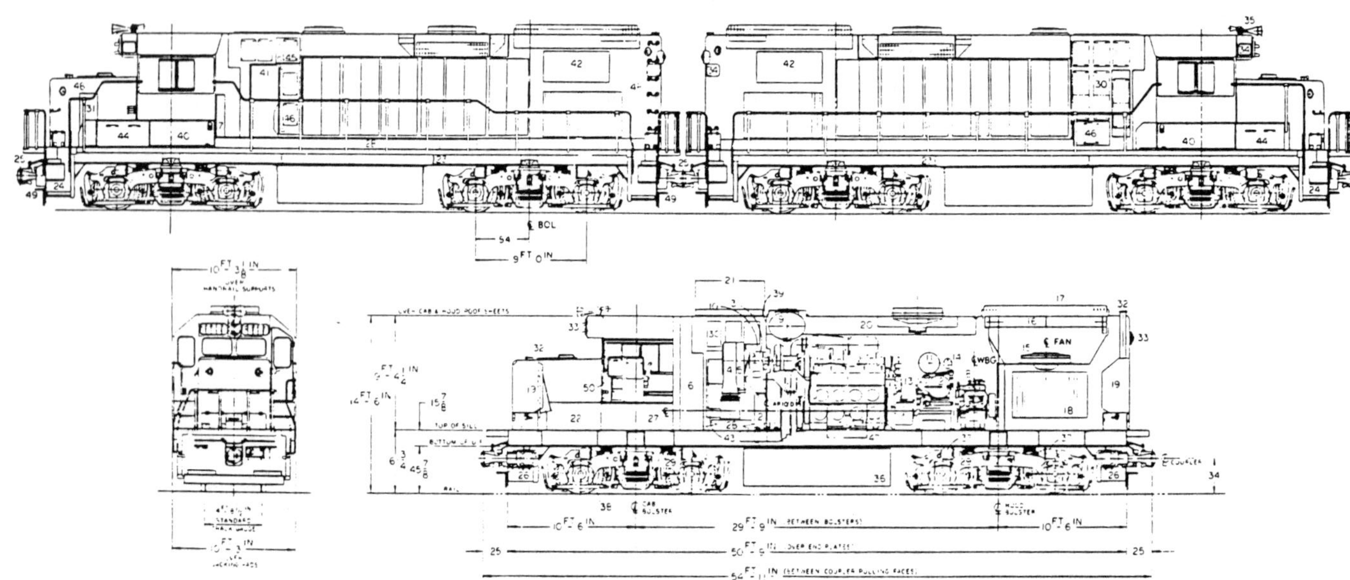

NOTE:
Locomotive Height Tolerance = ± 1½ in.
Locomotive Width Tolerance = ± ½ in.
Truck Lateral at Bolsters = ± 1¾ in. nominal
Locomotive is shown including half variable supplies and in new condition standing still on level and tangent track.

MODEL GP38-2 — 2000 HP GENERAL PURPOSE LOCOMOTIVE

LEGEND

1. Engine - 16-645E
2. Generator-Alternator AR10E1-D18
3. Traction Motor - Generator Blower
4. Auxiliary Generator
5. Electrical Control Cabinet
6. Air Compressor (WLN shown)
7. Engine Exhaust Stack
8. Handbrake
9. Sand Box Filler
10. Lube Oil Cooler
11. Engine Water Tank
12. 48 In. Cooling Fan "Q" Type
13. Radiator
14. Horn
15. Exhaust Manifold
16. Sand Box
*17. Fuel Filler
18. Headlight
19. Batteries
20. Fuel Tank (1700 Gal. shown)
21. Main Air Reservoirs
22. Cooling Air Inlet Shutters
23. Lube Oil Filter
24. Engine Air Filter
25. Inertial Air Separator
26. Number Box
27. Traction Motor Air Duct
28. Bell
*29. Dynamic Brake Fan
30. Pilot
31. Coupler
32. Fuel Filter
33. Truck - 4 Wheel
34. Electrical Cabinet Air Filter
35. Sand Trap Access
*36. M.U. Receptacle

*Modification

NOTE:
Locomotive Height Tolerance = ±1½ In.
Locomotive Width Tolerance = ±1/2 In.
Lateral (Bol. + Jrls. w/new trucks) = 2¾ In. Nom.

Locomotive is shown including half variable supplies and in new condition standing still on level and tangent track.

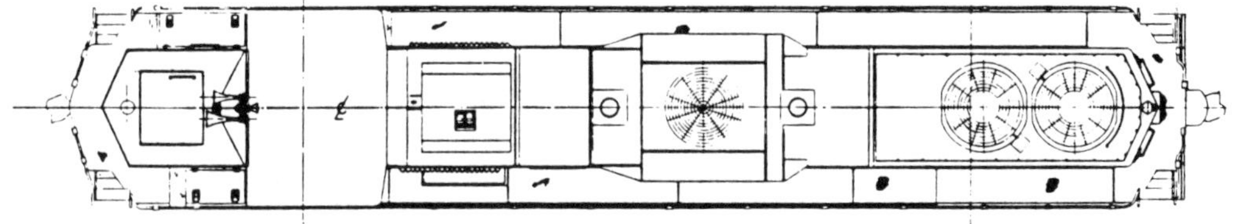

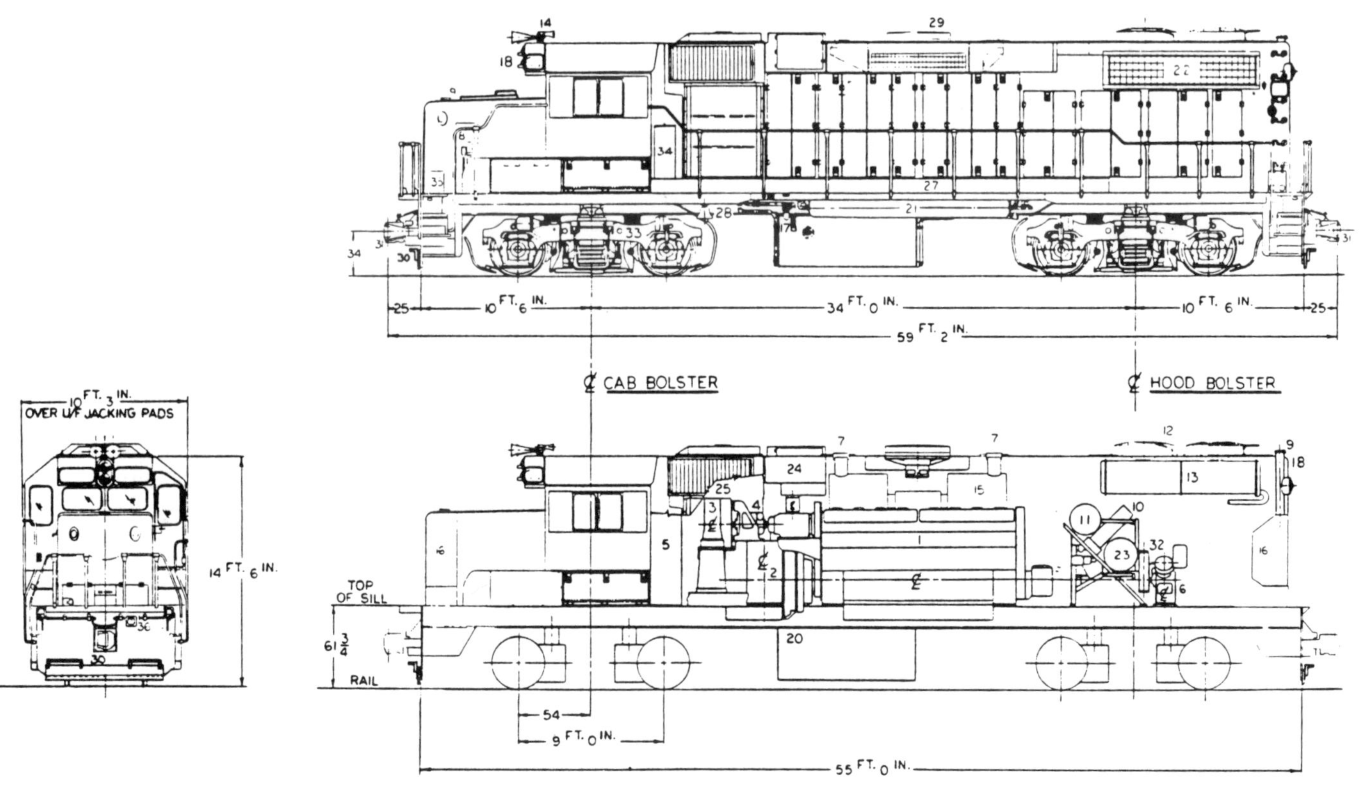

MODEL F40PH-2 — 3000 HP DIESEL ELECTRIC PASSENGER LOCOMOTIVE

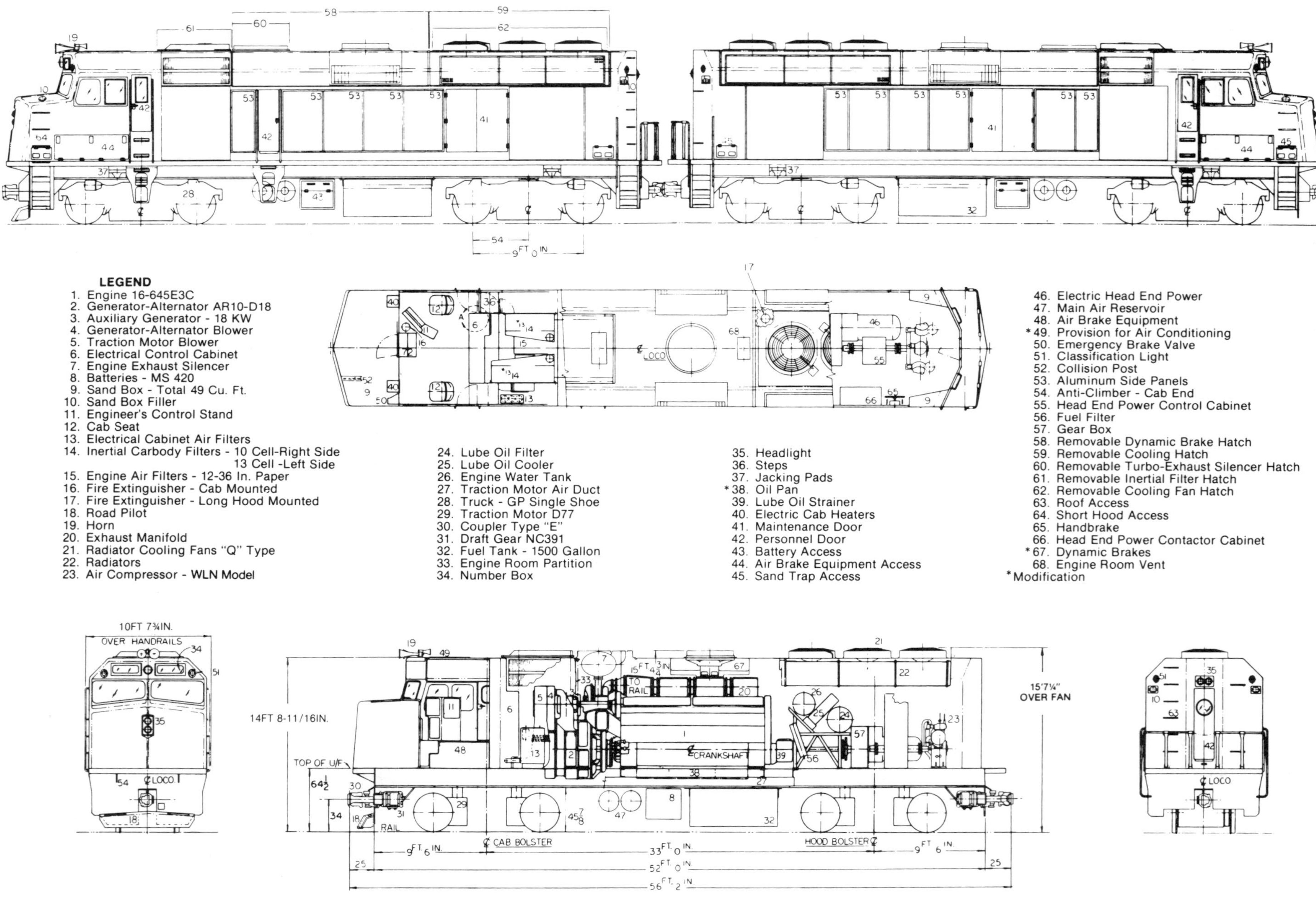

LEGEND

1. Engine 16-645E3C
2. Generator-Alternator AR10-D18
3. Auxiliary Generator - 18 KW
4. Generator-Alternator Blower
5. Traction Motor Blower
6. Electrical Control Cabinet
7. Engine Exhaust Silencer
8. Batteries - MS 420
9. Sand Box - Total 49 Cu. Ft.
10. Sand Box Filler
11. Engineer's Control Stand
12. Cab Seat
13. Electrical Cabinet Air Filters
14. Inertial Carbody Filters - 10 Cell-Right Side
    13 Cell -Left Side
15. Engine Air Filters - 12-36 In. Paper
16. Fire Extinguisher - Cab Mounted
17. Fire Extinguisher - Long Hood Mounted
18. Road Pilot
19. Horn
20. Exhaust Manifold
21. Radiator Cooling Fans "Q" Type
22. Radiators
23. Air Compressor - **WLN Model**
24. Lube Oil Filter
25. Lube Oil Cooler
26. Engine Water Tank
27. Traction Motor Air Duct
28. Truck - GP Single Shoe
29. Traction Motor D77
30. Coupler Type "E"
31. Draft Gear NC391
32. Fuel Tank - 1500 Gallon
33. Engine Room Partition
34. Number Box
35. Headlight
36. Steps
37. Jacking Pads
*38. Oil Pan
39. Lube Oil Strainer
40. Electric Cab Heaters
41. Maintenance Door
42. Personnel Door
43. Battery Access
44. Air Brake Equipment Access
45. Sand Trap Access
46. Electric Head End Power
47. Main Air Reservoir
48. Air Brake Equipment
*49. Provision for Air Conditioning
50. Emergency Brake Valve
51. Classification Light
52. Collision Post
53. Aluminum Side Panels
54. Anti-Climber - Cab End
55. Head End Power Control Cabinet
56. Fuel Filter
57. Gear Box
58. Removable Dynamic Brake Hatch
59. Removable Cooling Hatch
60. Removable Turbo-Exhaust Silencer Hatch
61. Removable Inertial Filter Hatch
62. Removable Cooling Fan Hatch
63. Roof Access
64. Short Hood Access
65. Handbrake
66. Head End Power Contactor Cabinet
*67. Dynamic Brakes
68. Engine Room Vent

*Modification

MODEL F40PH-2 CAB ARRANGEMENT

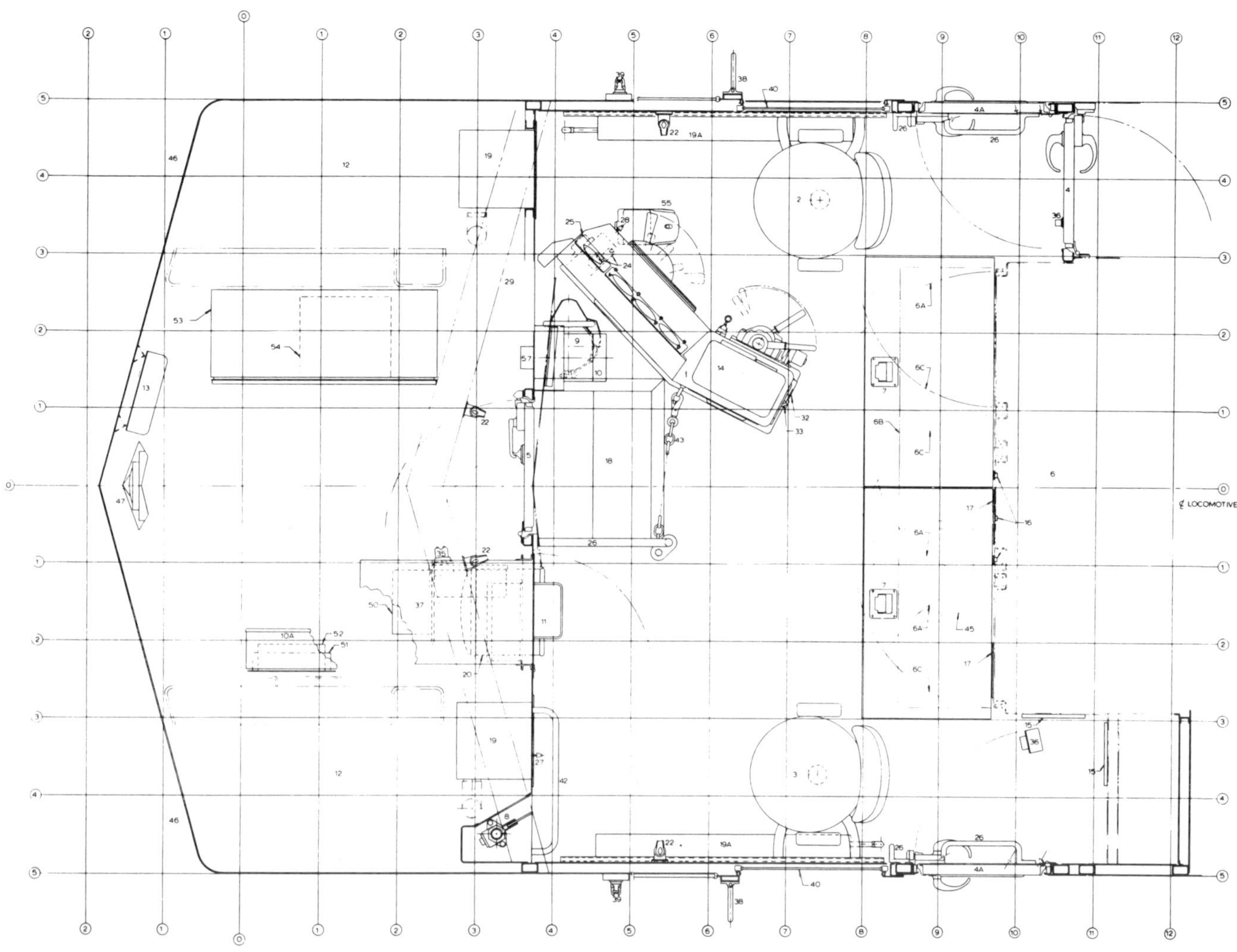

LEGEND

1. ENGINEER'S CONTROL STAND
2. ENGINEER'S SLIDING SEAT
3. OBSERVER'S SLIDING SEAT
4. CAB DOOR

4A. CAB DOOR WITH SLIDING WINDOWS

5. CAB SHORT HOOD DOOR
6. ELECTRICAL CABINET

6A. UPPER DOOR

6B. MIDDLE DOOR

6C. LOWER DOOR

7. TRAP (2)
8. EMERGENCY BRAKE VALVE
9. FIRE EXTINGUISHER
10. SPEED INDICATOR

10A. SPEED RECORDER

*11. WASTE RECEPTACLE

12. SAND BOX (2)
13. FUSEE & TORPEDO BOX

*14. TRAIN COMMUNICATION EQUIPMENT

15. CARD HOLDER (2)
16. FOLDING COAT HOOK (3)
17. INSTRUCTION PLATE (2)
18. STEPS
19. CAB HEATER - FRONT (2)

19A. CAB HEATER - STRIP (2)

*20. STANDBY CAB HEATER

22. SUN VISOR (4)
23. WINDOW WIPER AND GUARD
24. BRAKE AND REVERSER HANDLE HOLDER

*25. WHEEL SLIP BUZZER

26. GRAB IRON (5)
27. SWITCH - STRIP HEATER (LEFT SIDE)
28. SWITCH - STRIP HEATER (RIGHT SIDE)
29. TOP OF DASHBOARD
30. DEFROSTER DUCT (2)
31. NUMBER BOX
32. MU2A VALVE
33. FEED VALVE
34. WINDSHIELD
35. DOOR STOP
36. DOOR BUMPER (2)

*37. RADIO EQUIPMENT BOX

*38. WIND DEFLECTOR WITH REAR VIEW MIRROR (2)

39. LAMP BRACKET (2)
40. SIDE SASH UNIT (2)

*41. SIGNAL LIGHT CONTROL SWITCH

42. GUARD
43. SAFETY CHAIN

*44. SAFETY CONTROL HORN BOX

45. CAB ROOF VENTILATOR
47. TWIN SEALED BEAM (BOTH ENDS OF UNIT)

*50. TRANS VOLTAGE FILTER

*51. CONTROL BOX ASM

*52. POWER CONVERTER

*53. EQUIPMENT BOX

*54. CONVERTOR

*55. SWITCH FOOT ACK.

*56. CAB SIGNAL - AUTOMATIC TRAIN CONTROL

*57. TIMING GAUGE

*MODIFICATIONS

MODEL GP40-2 — 3000 HP DIESEL ELECTRIC GENERAL PURPOSE LOCOMOTIVE

LEGEND

1. ENGINE - 16-645E3C
2. GENERATOR/ALTERNATOR AR10A6/D18
3. TRACTION MOTOR - GENERATOR BLOWER
4. AUXILIARY GENERATOR
5. ELECTRICAL CONTROL CABINET.
6. AIR COMPRESSOR (WBO SHOWN)
7. ENGINE EXHAUST SILENCER
8. HANDBRAKE
9. SAND BOX FILLER
10. LUBE OIL COOLER
11. ENGINE WATER TANK
12. 48 IN. COOLING FAN "Q" TYPE
13. RADIATOR
14. HORN
15. EXHAUST MANIFOLD
16. SAND BOX
*17. MU END CONNECTION
18. HEADLIGHT
19. BATTERIES
20. FUEL TANK (2600 GAL. SHOWN)
21. MAIN AIR RESERVOIRS
22. COOLING AIR INLET & SHUTTERS
23. LUBE OIL FILTER
24. ENGINE AIR FILTER
25. INERTIAL AIR SEPARATOR
26. NUMBER BOX
27. TRACTION MOTOR AIR DUCT
28. BELL
*29. DYNAMIC BRAKE FAN
30. PILOT
31. COUPLER
32. FUEL FILTER
33. TRUCK - 4 WHEEL
34. ELECTRICAL CABINET AIR FILTER
35. ENGINE ROOM VENT

*MODIFICATION

NOTE:
LOCOMOTIVE HEIGHT TOLERANCE = ±1½ IN.
LOCOMOTIVE WIDTH TOLERANCE = ±1/2 IN.
LATERAL (Bol. + Jrls. w/new trucks) = ±2¾ IN. NOM.

Locomotive is shown including half variable supplies and in new condition standing still on level and tangent track.

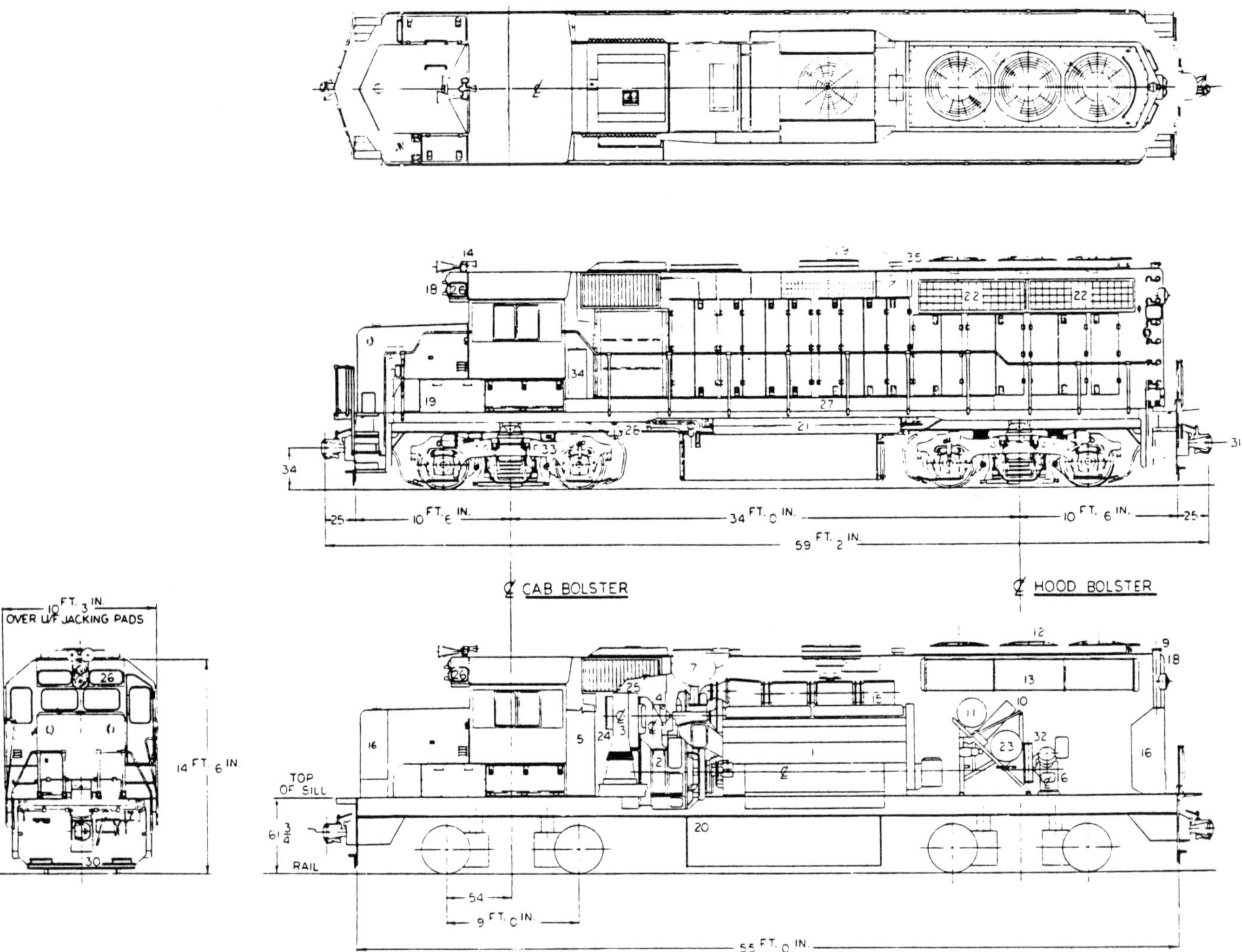

MODEL SD40-2 — 3000 HP SIX MOTOR DIESEL ELECTRIC LOCOMOTIVE

LEGEND

1. ENGINE - 16-645E3C
2. GENERATOR/ALTERNATOR AR10/D18
3. TRACTION MOTOR BLOWER
4. GENERATOR/ALTERNATOR BLOWER
*5. AUXILIARY GENERATOR (18 KW)
6. AIR COMPRESSOR (WLN SHOWN)
7. ELECTRICAL CONTROL CABINET
8. HANDBRAKE (WHEEL TYPE)
9. SAND BOX FILLER
10. LUBE OIL COOLER
11. ENGINE WATER TANK
12. COOLING FAN (48"-"Q" TYPE - 3 TOTAL)
13. RADIATOR
14. HORN
15. EXHAUST MANIFOLD
16. SAND BOX
*17. FUEL FILLER
18. HEADLIGHT
19. BATTERY BOX ACCESS
*20. FUEL TANK (4000 GAL. SHOWN)
21. MAIN AIR RESERVOIR
22. COOLING AIR INLET & SHUTTERS
23. LUBE OIL FILTER
24. ENGINE AIR FILTER
25. INERTIAL AIR FILTER
26. NUMBER BOX
27. TRACTION MOTOR AIR DUCT
28. BELL
*29. DYNAMIC BRAKE FAN (48 IN.)
30. PILOT
31. COUPLER
32. FUEL FILTER
33. TRUCK HTC-3 AXLE
34. ELECTRICAL CABINET AIR FILTER BOX
35. ENGINE ROOM VENT
36. ENGINE EXHAUST SILENCER
37. TRACTION MOTOR
38. DUST BIN BLOWER
39. INERTIAL FILTER AIR INLET
*40. DYNAMIC BRAKE GRIDS
*41. DYNAMIC BRAKE AIR INLET
*42. DYNAMIC BRAKE CABLE ACCESS
*43. "D" CONTACTOR ACCESS
44. ENGINE ROOM PARTITION
45. FUEL PUMP
46. SAND TRAP ACCESS
47. LADDER FOR ROOF ACCESS
48. AIR BRAKE EQUIPMENT ACCESS
49. REMOVABLE FILTER HATCH
50. REMOVABLE TURBO HATCH
*51. REMOVABLE DYNAMIC BRAKE HATCH
52. REMOVABLE FAN HATCH
53. HOOD LATERAL STABILIZERS
54. CAB DOOR
55. ENGINE ROOM ACCESS DOORS

*MODIFICATIONS

NOTE:
LOCOMOTIVE HEIGHT TOLERANCE = +2½ IN./-1½ IN.
LOCOMOTIVE WIDTH TOLERANCE = ±1/2 IN.
LATERAL (Bol. + Jrls. w/new trucks) = 2¼ IN. NOM.

Locomotive is shown including half variable supplies and in new condition standing still on level and tangent track.

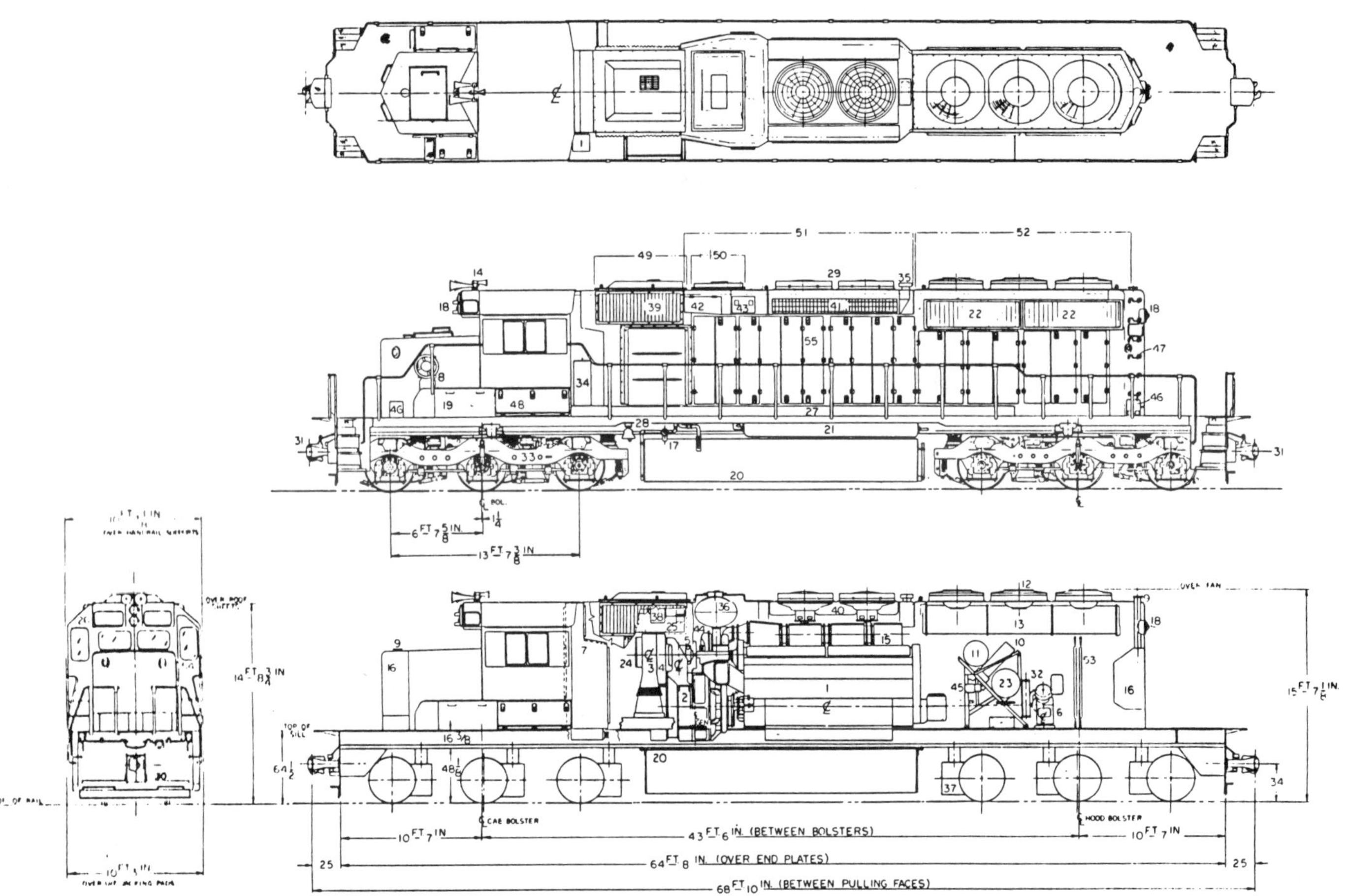

MODEL SD40-2 DELUXE CANADIAN CAB – 3000 HP SIX MOTOR DIESEL-ELECTRIC LOCOMOTIVE

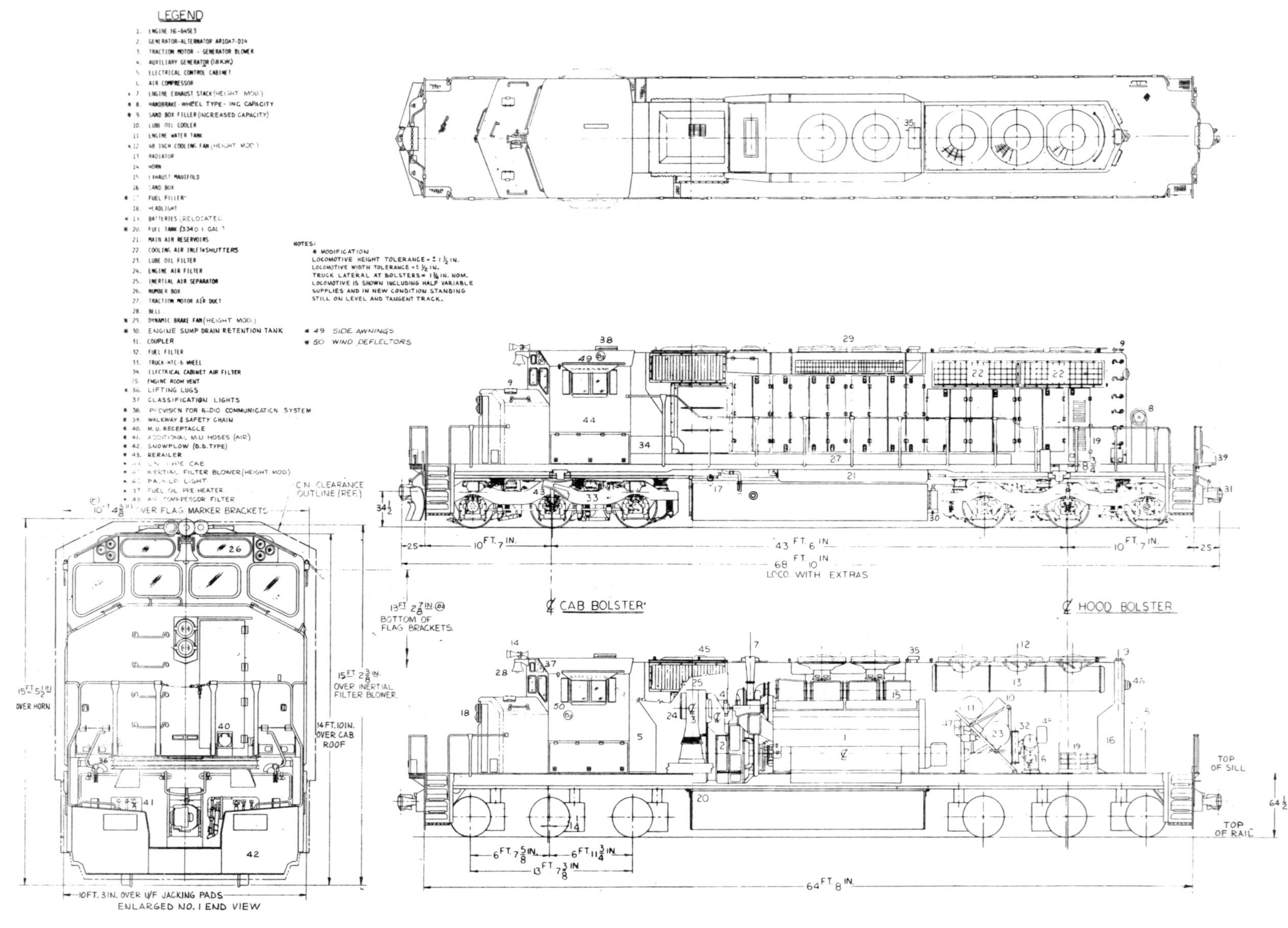

Elevation Drawings of the DD-GMCL. Deluxe Canadian Cab. SD40-2 DCC Model.

MODEL GP50 — 3500 HP
GENERAL PURPOSE LOCOMOTIVE

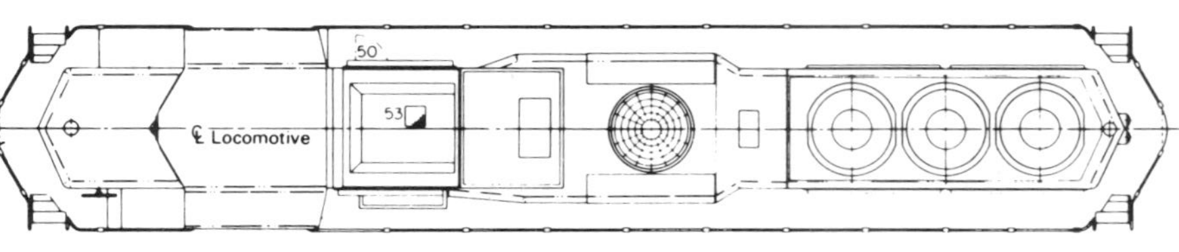

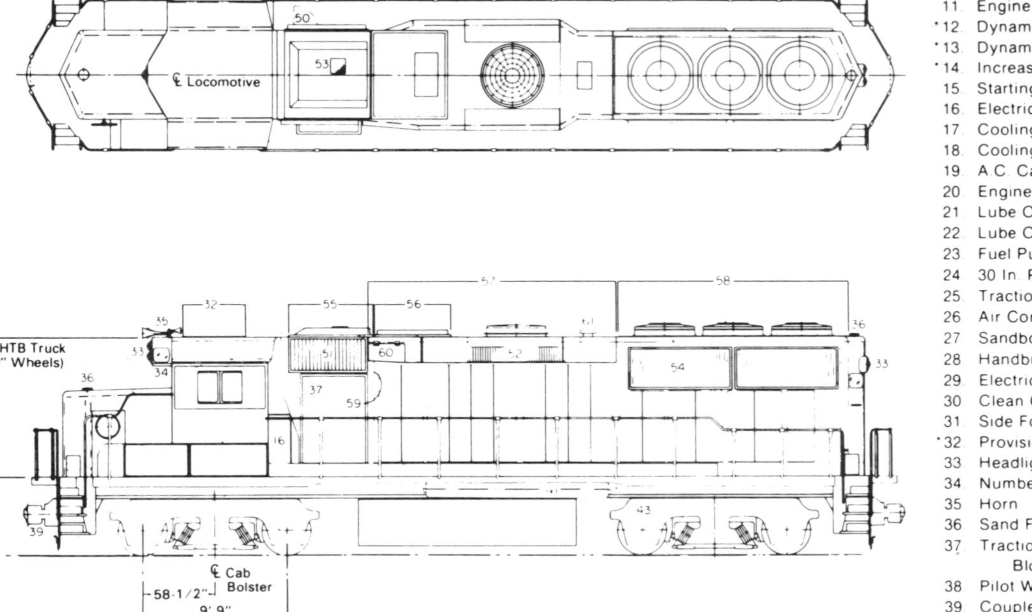

LEGEND

1. Engine 16-645F3B
2. Generator Alternator (AR15 D18)
3. Auxiliary Generator (18 KW-AC)
4. Traction Motor Blower
5. Generator/Alternator Blower
6. Engine Exhaust Silencer
7. Engine Air Filters
8. Electrical Control Cabinet
9. Inertial Filters
10. Dust Bin Blower
11. Engine Room Partition
*12. Dynamic Brake Fan
*13. Dynamic Brake Grids
*14. Increased Capacity Oil Pan
15. Starting Motors
16. Electrical Control Cabinet Air Filter Box
17. Cooling Radiators
18. Cooling Fans
19. A.C. Cabinet
20. Engine Water Tank
21. Lube Oil Cooler
22. Lube Oil Filter
23. Fuel Pump
24. 30 In. Primary Air Filter
25. Traction Motor Air Duct
26. Air Compressor (WBO)
27. Sandbox
28. Handbrake
29. Electric Cab Heater
30. Clean Cab (Phase II)
31. Side Footboards (4 Corners)
*32. Provision for Air Conditioning
33. Headlight
34. Number Boxes
35. Horn
36. Sand Filler
37. Traction Motor - Generator/Alternator Blower Blower Enclosure
38. Pilot W/Hose Storage Rack
39. Coupler
40. Draft Gear
41. Main Reservoirs
42. GP Truck (2 Axle - 40" Diameter Wheels)
*43. HTB Truck (2 Axle - 40" Diameter Wheels)
44. Traction Motors
45. Fuel Tank 3000 Gal.
46. Sand Trap Access
47. Ladder for Roof Access
48. Battery Access
49. Air Brake Equipment Access
50. Inertial Filter Compartment Access Door
51. Inertial Filter Air Inlet
*52. Dynamic Brake Air Inlet
53. Dust Bin Blower Air Discharge
54. Cooling Fan Inlet Shutters
55. Removable Filter Hatch
56. Removable Turbo Hatch
*57. Removable Dynamic Brake Hatch
58. Removable Fan Hatch
59. Bolted Joint
*60. Dynamic Brake Cable Connection Access
61. Engine Room Vent
62. Hood Lateral Stabilizers

* Modification

Note:

Locomotive Height Tolerance = ± 1-1/2 In.
Locomotive Width Tolerance = ± 1/2 In.
Truck Lateral at Bolsters = ± 1-3/4 In. Nom.

Locomotive is shown including half variable supplies and in new condition standing still on level and tangent track.

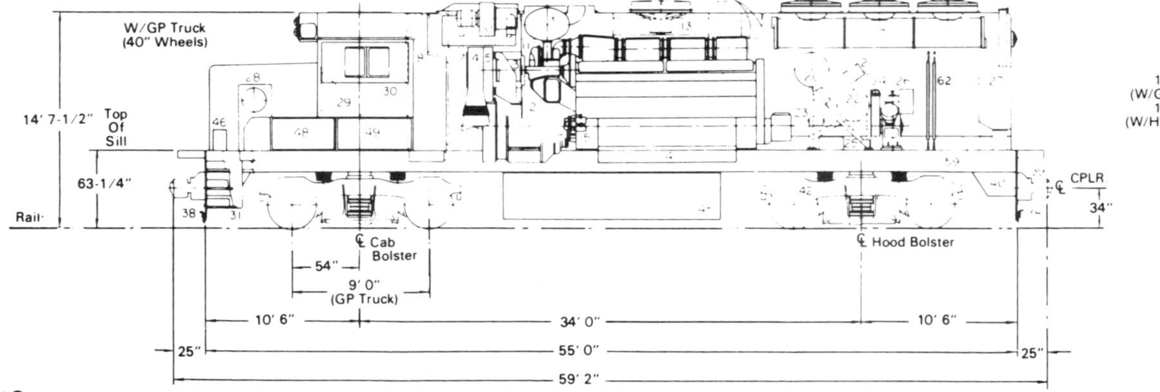

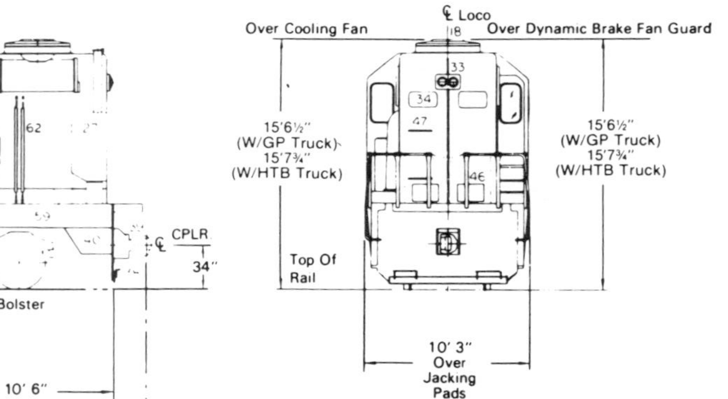

MODEL SD50 - 3500 HP
SIX MOTOR DIESEL-ELECTRIC LOCOMOTIVE

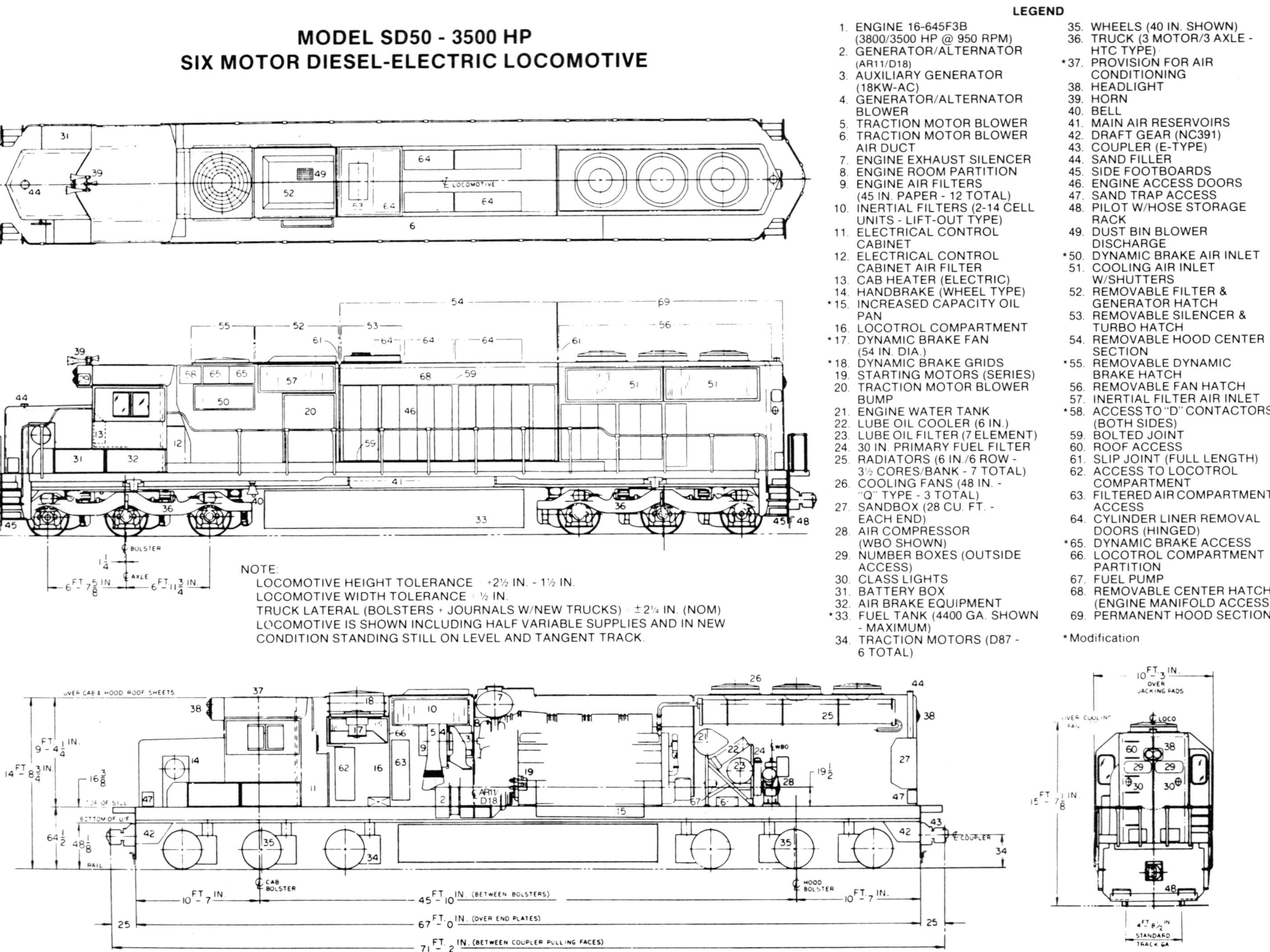

LEGEND

1. ENGINE 16-645F3B (3800/3500 HP @ 950 RPM)
2. GENERATOR/ALTERNATOR (AR11/D18)
3. AUXILIARY GENERATOR (18KW-AC)
4. GENERATOR/ALTERNATOR BLOWER
5. TRACTION MOTOR BLOWER
6. TRACTION MOTOR BLOWER AIR DUCT
7. ENGINE EXHAUST SILENCER
8. ENGINE ROOM PARTITION
9. ENGINE AIR FILTERS (45 IN. PAPER - 12 TOTAL)
10. INERTIAL FILTERS (2-14 CELL UNITS - LIFT-OUT TYPE)
11. ELECTRICAL CONTROL CABINET
12. ELECTRICAL CONTROL CABINET AIR FILTER
13. CAB HEATER (ELECTRIC)
14. HANDBRAKE (WHEEL TYPE)
*15. INCREASED CAPACITY OIL PAN
16. LOCOTROL COMPARTMENT
*17. DYNAMIC BRAKE FAN (54 IN. DIA.)
*18. DYNAMIC BRAKE GRIDS
19. STARTING MOTORS (SERIES)
20. TRACTION MOTOR BLOWER BUMP
21. ENGINE WATER TANK
22. LUBE OIL COOLER (6 IN.)
23. LUBE OIL FILTER (7 ELEMENT)
24. 30 IN. PRIMARY FUEL FILTER
25. RADIATORS (6 IN./6 ROW - 3½ CORES/BANK - 7 TOTAL)
26. COOLING FANS (48 IN. - "Q" TYPE - 3 TOTAL)
27. SANDBOX (28 CU. FT. - EACH END)
28. AIR COMPRESSOR (WBO SHOWN)
29. NUMBER BOXES (OUTSIDE ACCESS)
30. CLASS LIGHTS
31. BATTERY BOX
32. AIR BRAKE EQUIPMENT
*33. FUEL TANK (4400 GA. SHOWN - MAXIMUM)
34. TRACTION MOTORS (D87 - 6 TOTAL)
35. WHEELS (40 IN. SHOWN)
36. TRUCK (3 MOTOR/3 AXLE - HTC TYPE)
*37. PROVISION FOR AIR CONDITIONING
38. HEADLIGHT
39. HORN
40. BELL
41. MAIN AIR RESERVOIRS
42. DRAFT GEAR (NC391)
43. COUPLER (E-TYPE)
44. SAND FILLER
45. SIDE FOOTBOARDS
46. ENGINE ACCESS DOORS
47. SAND TRAP ACCESS
48. PILOT W/HOSE STORAGE RACK
49. DUST BIN BLOWER DISCHARGE
*50. DYNAMIC BRAKE AIR INLET
51. COOLING AIR INLET W/SHUTTERS
52. REMOVABLE FILTER & GENERATOR HATCH
53. REMOVABLE SILENCER & TURBO HATCH
54. REMOVABLE HOOD CENTER SECTION
*55. REMOVABLE DYNAMIC BRAKE HATCH
56. REMOVABLE FAN HATCH
57. INERTIAL FILTER AIR INLET
*58. ACCESS TO "D" CONTACTORS (BOTH SIDES)
59. BOLTED JOINT
60. ROOF ACCESS
61. SLIP JOINT (FULL LENGTH)
62. ACCESS TO LOCOTROL COMPARTMENT
63. FILTERED AIR COMPARTMENT ACCESS
64. CYLINDER LINER REMOVAL DOORS (HINGED)
*65. DYNAMIC BRAKE ACCESS
66. LOCOTROL COMPARTMENT PARTITION
67. FUEL PUMP
68. REMOVABLE CENTER HATCH (ENGINE MANIFOLD ACCESS)
69. PERMANENT HOOD SECTION

*Modification

NOTE:
LOCOMOTIVE HEIGHT TOLERANCE = +2½ IN. - 1½ IN.
LOCOMOTIVE WIDTH TOLERANCE = ½ IN.
TRUCK LATERAL (BOLSTERS + JOURNALS W/NEW TRUCKS) = ±2¼ IN. (NOM)
LOCOMOTIVE IS SHOWN INCLUDING HALF VARIABLE SUPPLIES AND IN NEW CONDITION STANDING STILL ON LEVEL AND TANGENT TRACK.

MODEL GP59 – 3000 HP Four Motor Diesel-Electric Locomotive

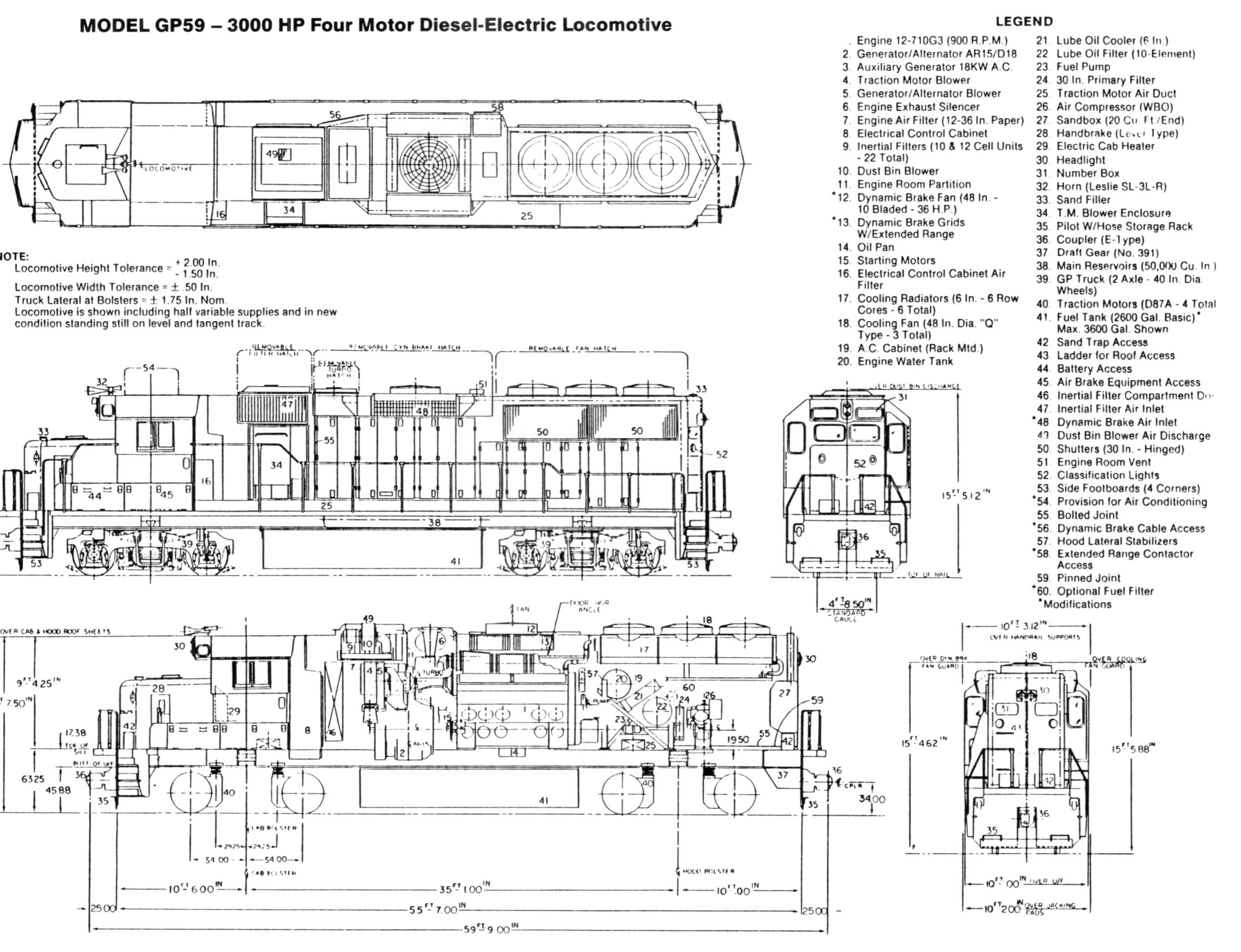

NOTE:
Locomotive Height Tolerance = + 2.00 In. / - 1.50 In.
Locomotive Width Tolerance = ± .50 In.
Truck Lateral at Bolsters = ± 1.75 In. Nom.
Locomotive is shown including half variable supplies and in new condition standing still on level and tangent track.

LEGEND

. Engine 12-710G3 (900 R.P.M.)
2. Generator/Alternator AR15/D18
3. Auxiliary Generator 18KW A.C.
4. Traction Motor Blower
5. Generator/Alternator Blower
6. Engine Exhaust Silencer
7. Engine Air Filter (12-36 In. Paper)
8. Electrical Control Cabinet
9. Inertial Filters (10 & 12 Cell Units - 22 Total)
10. Dust Bin Blower
11. Engine Room Partition
*12. Dynamic Brake Fan (48 In. - 10 Bladed - 36 H.P.)
*13. Dynamic Brake Grids W/Extended Range
14. Oil Pan
15. Starting Motors
16. Electrical Control Cabinet Air Filter
17. Cooling Radiators (6 In. - 6 Row Cores - 6 Total)
18. Cooling Fan (48 In. Dia. "Q" Type - 3 Total)
19. A.C. Cabinet (Rack Mtd.)
20. Engine Water Tank
21. Lube Oil Cooler (6 In.)
22. Lube Oil Filter (10-Element)
23. Fuel Pump
24. 30 In. Primary Filter
25. Traction Motor Air Duct
26. Air Compressor (WBO)
27. Sandbox (20 Cu. Ft./End)
28. Handbrake (Lever Type)
29. Electric Cab Heater
30. Headlight
31. Number Box
32. Horn (Leslie SL-3L-R)
33. Sand Filler
34. T.M. Blower Enclosure
35. Pilot W/Hose Storage Rack
36. Coupler (E-Type)
37. Draft Gear (No. 391)
38. Main Reservoirs (50,000 Cu. In.)
39. GP Truck (2 Axle - 40 In. Dia. Wheels)
40. Traction Motors (D87A - 4 Total)
41. Fuel Tank (2600 Gal. Basic)* Max. 3600 Gal. Shown
42. Sand Trap Access
43. Ladder for Roof Access
44. Battery Access
45. Air Brake Equipment Access
46. Inertial Filter Compartment Do
47. Inertial Filter Air Inlet
*48. Dynamic Brake Air Inlet
49. Dust Bin Blower Air Discharge
50. Shutters (30 In. - Hinged)
51. Engine Room Vent
52. Classification Lights
53. Side Footboards (4 Corners)
*54. Provision for Air Conditioning
55. Bolted Joint
*56. Dynamic Brake Cable Access
57. Hood Lateral Stabilizers
*58. Extended Range Contactor Access
59. Pinned Joint
*60. Optional Fuel Filter

*Modifications

MODEL GP60 - 3800 HP
FOUR MOTOR DIESEL-ELECTRIC LOCOMOTIVE

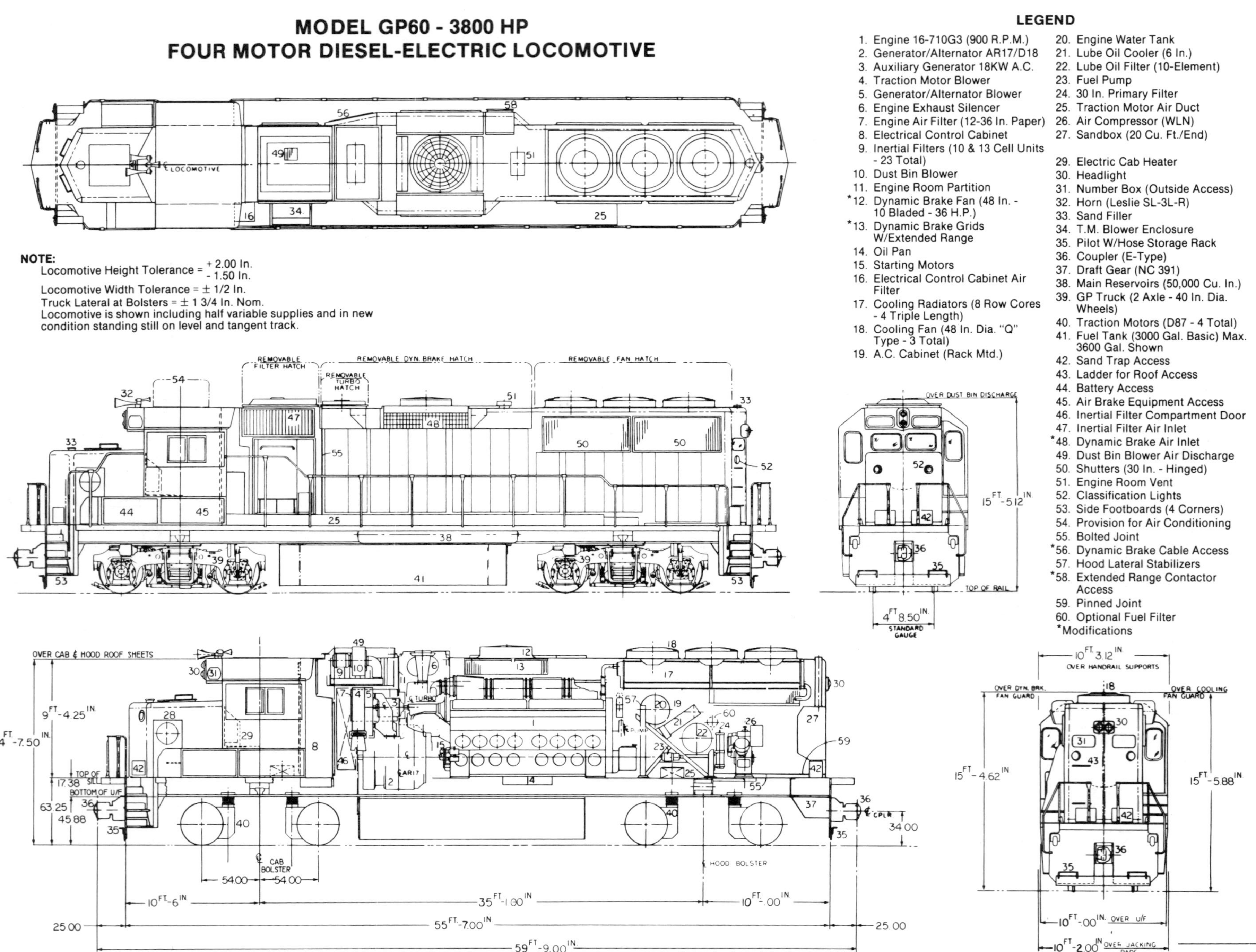

NOTE:

Locomotive Height Tolerance = + 2.00 In. / - 1.50 In.

Locomotive Width Tolerance = ± 1/2 In.
Truck Lateral at Bolsters = ± 1 3/4 In. Nom.
Locomotive is shown including half variable supplies and in new condition standing still on level and tangent track.

LEGEND

1. Engine 16-710G3 (900 R.P.M.)
2. Generator/Alternator AR17/D18
3. Auxiliary Generator 18KW A.C.
4. Traction Motor Blower
5. Generator/Alternator Blower
6. Engine Exhaust Silencer
7. Engine Air Filter (12-36 In. Paper)
8. Electrical Control Cabinet
9. Inertial Filters (10 & 13 Cell Units - 23 Total)
10. Dust Bin Blower
11. Engine Room Partition
*12. Dynamic Brake Fan (48 In. - 10 Bladed - 36 H.P.)
*13. Dynamic Brake Grids W/Extended Range
14. Oil Pan
15. Starting Motors
16. Electrical Control Cabinet Air Filter
17. Cooling Radiators (8 Row Cores - 4 Triple Length)
18. Cooling Fan (48 In. Dia. "Q" Type - 3 Total)
19. A.C. Cabinet (Rack Mtd.)
20. Engine Water Tank
21. Lube Oil Cooler (6 In.)
22. Lube Oil Filter (10-Element)
23. Fuel Pump
24. 30 In. Primary Filter
25. Traction Motor Air Duct
26. Air Compressor (WLN)
27. Sandbox (20 Cu. Ft./End)
29. Electric Cab Heater
30. Headlight
31. Number Box (Outside Access)
32. Horn (Leslie SL-3L-R)
33. Sand Filler
34. T.M. Blower Enclosure
35. Pilot W/Hose Storage Rack
36. Coupler (E-Type)
37. Draft Gear (NC 391)
38. Main Reservoirs (50,000 Cu. In.)
39. GP Truck (2 Axle - 40 In. Dia. Wheels)
40. Traction Motors (D87 - 4 Total)
41. Fuel Tank (3000 Gal. Basic) Max. 3600 Gal. Shown
42. Sand Trap Access
43. Ladder for Roof Access
44. Battery Access
45. Air Brake Equipment Access
46. Inertial Filter Compartment Door
47. Inertial Filter Air Inlet
*48. Dynamic Brake Air Inlet
49. Dust Bin Blower Air Discharge
50. Shutters (30 In. - Hinged)
51. Engine Room Vent
52. Classification Lights
53. Side Footboards (4 Corners)
54. Provision for Air Conditioning
55. Bolted Joint
*56. Dynamic Brake Cable Access
57. Hood Lateral Stabilizers
*58. Extended Range Contactor Access
59. Pinned Joint
60. Optional Fuel Filter

*Modifications

MODEL SD60 – 3800 HP
SIX MOTOR DIESEL-ELECTRIC LOCOMOTIVE

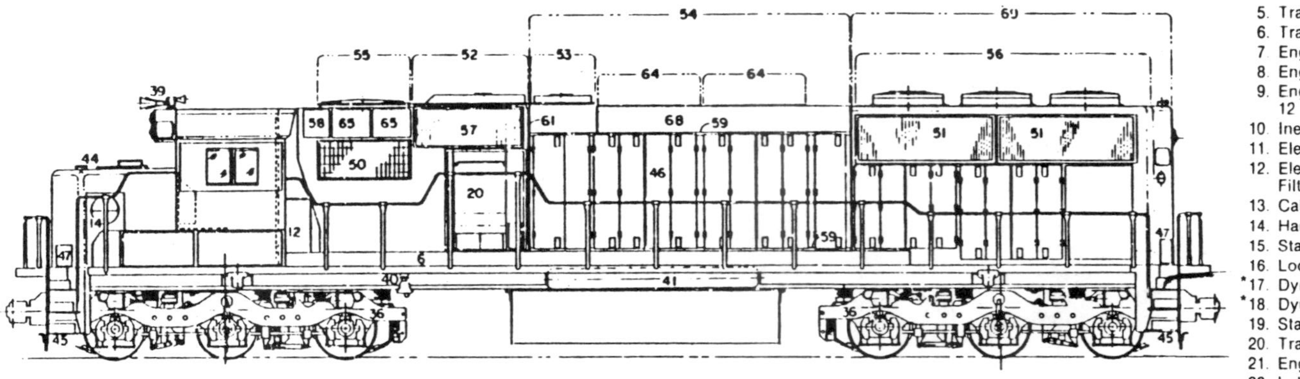

LEGEND

1. Engine 16-710G3 (3950/3800 HP @ 900 RMP)
2. Generator/Alternator (AR11/D18A)
3. Auxiliary Generator (18kW-AC)
4. Generator/Alternator Blower
5. Traction Motor Blower
6. Traction Motor Blower Air Duct
7. Engine Exhaust Silencer
8. Engine Room Partition
9. Engine Air Filters (45 In. Paper - 12 Total)
10. Inertial Filters (2-14 Cell Units)
11. Electrical Control Cabinet
12. Electrical Control Cabinet Air Filter
13. Cab Heater (Electric)
14. Handbrake (Wheel Type)
15. Standard Oil Pan
16. Locotrol Compartment
*17. Dynamic Brake Fan (54 In. Dia.)
*18. Dynamic Brake Grids
19. Starting Motors
20. Traction Motor Blower Pump
21. Engine Water Tank
22. Lube Oil Cooler (6 In.)
23. Lube Oil Filter (10 Element)
24. 30 In. Primary Fuel Filter (2 Total)
25. Radiators (6 In./6 Row - 3-1/2 Cores/Bank
26. Cooling Fans (48 In. Dia. - 2 Speed - 4 Blade "Q" Type - 3 Total)
27. Sandbox (28 Cu. Ft. - Each End)
28. Air Compressor (WLN Shown)
29. Number Boxes (Outside Access)
30. Class Lights
31. Battery Box
32. Air Brake Equipment
33. Fuel Tank (3200 Gal. Shown)
34. Traction Motor (D87B - 6 Total)
35. Wheels (40 In. Shown)
36. Truck (3 Motor/3 Axle - HTC Type)
*37. Provision for Air Conditioning
38. Headlight
39. Horn
40. Bell
41. Main Air Reservoirs
42. Draft Gear (NC391)
43. Coupler (E-Type)
44. Sand Filler
45. Side Footboards
46. Engine Access Doors
47. Sand Trap Access
48. Pilot w/Hose Storage Rack
49. Dust Bin Blower Discharge
*50. Dynamic Brake Air Inlet
51. Cooling Air Inlet w/Shutters
52. Removable Filter & Generator Hatch
53. Removable Silencer & Turbo Hatch
54. Removable Hood Center Section
55. Removable Dynamic Brake Hatch
56. Removable Fan Hatch
57. Inertial Filter Air Inlet
*58. Access to "D" Contactors (Both Sides)
59. Bolted Joint
60. Roof Access
61. Slip Joint (Full Length)
62. Access to Locotrol Compartment
63. Filtered Air Compartment Access
64. Cylinder Liner Removal Doors (Hinged)
*65. Dynamic Brake Access
66. Locotrol Compartment Partition
67. Fuel Pump
68. Removable Center Hatch (Engine Manifold Access)
69. Permanent Hood End Section
70. Air Compressor Clutch
71. Traction Motor Blower Inlet Shutter

*Modification

NOTE:
LOCOMOTIVE HEIGHT TOLERANCE = +2-1/2 IN., -1-1/2 IN.
LOCOMOTIVE WIDTH TOLERANCE = 1/2 IN.
TRUCK LATERAL (BOLSTERS + JOURNALS W/NEW TRUCK = ±2-1/4 IN. NOM.

LOCOMOTIVE IS SHOWN INCLUDING HALF VARIABLES SUPPLIES AND IN NEW CONDITION STANDING STILL ON LEVEL AND TANGENT TRACK.

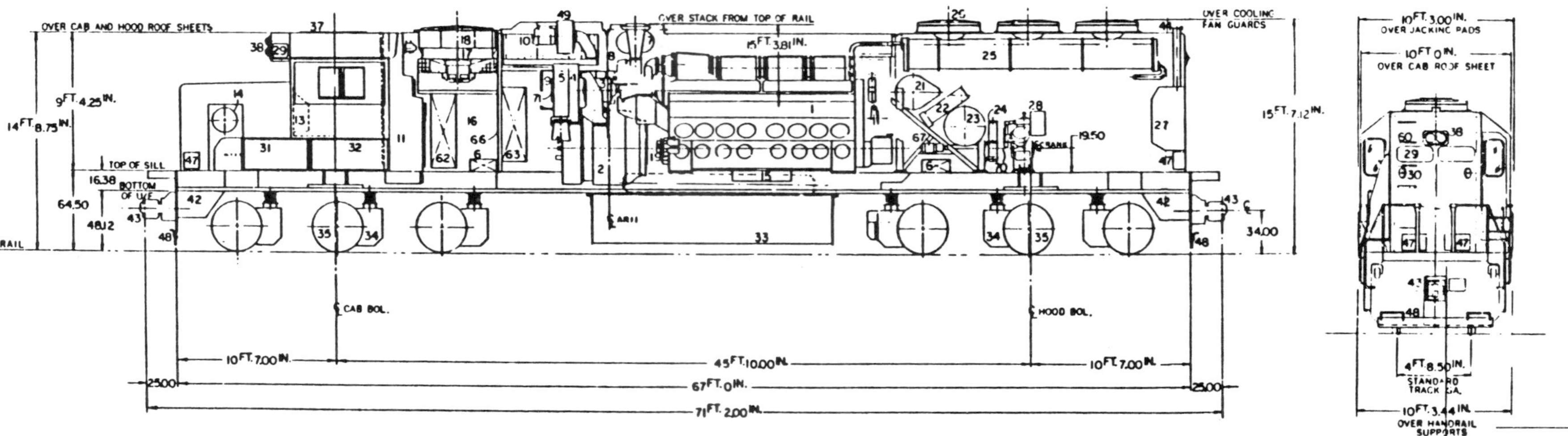

MODEL SD60 – 3800 HP Diesel-Electric Locomotive

CAB AND SHORT HOOD ARRANGEMENT

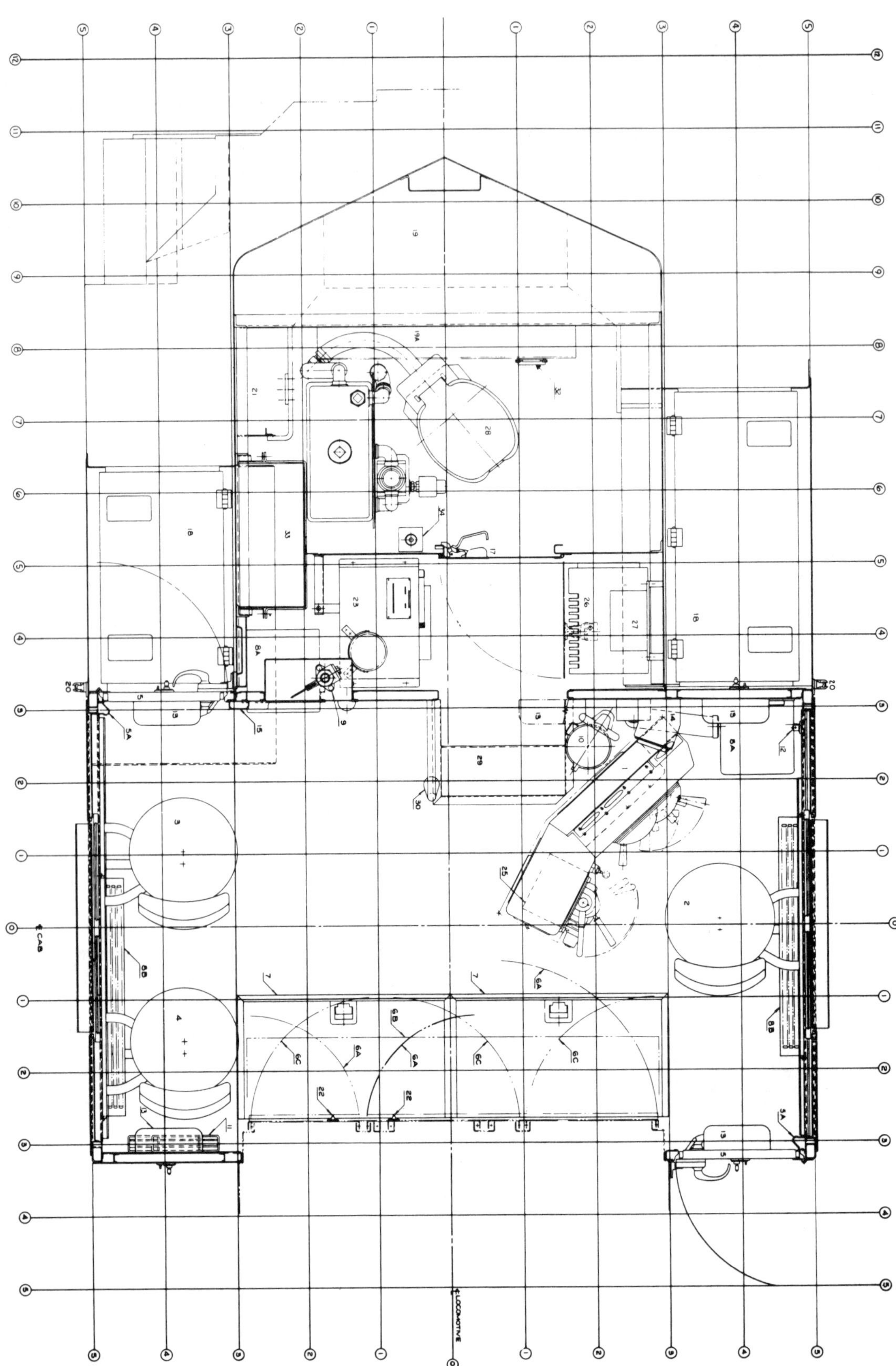

1. Engineer's Control Stand - AAR
2. Engineer's Seat with Seat Locking Modification
*3. Auxiliary Seat
4. Observer's Seat
5. Cab Door (2)
5A. Door Hinge Boot (2)
6. Electrical Cabinet
6A. Upper Doors
6B. Middle Door
6C. Lower Doors
7. Trap Doors (2)
8. Cab Heaters - Electric (4)
8A. Front Partition (2)
8B. Side Strip Heater (2)
9. Emergency Brake Valve & Guard
*10. Fire Extinguisher - 30 Lb.
11. Flag & Fusee Rack
12. Sun Visor
13. Window Wiper & Guard (6)
14. Speed Recorder
15. Defroster Duct
16. Door Stop
17. Short Hood Door
18. Battery Box
19. Sand Box
*19A. Extra Capacity Sand Box
20. Lamp Bracket (2)
21. Hand Brake Recess
22. Coat Hook (2)
*23. Water Cooler
24. Card Holder
*25. Radio
*26. Transmitter & Receiver
*27. Radio Voltage Filter
*28. Toilet
29. Step Down to Short Hood
30. Grab Iron
31. Headbump Pad
*32. Toilet Paper Holder
*33. Water Tank
34. Floor Drain

MODEL SD60 CLEARANCE DIAGRAM

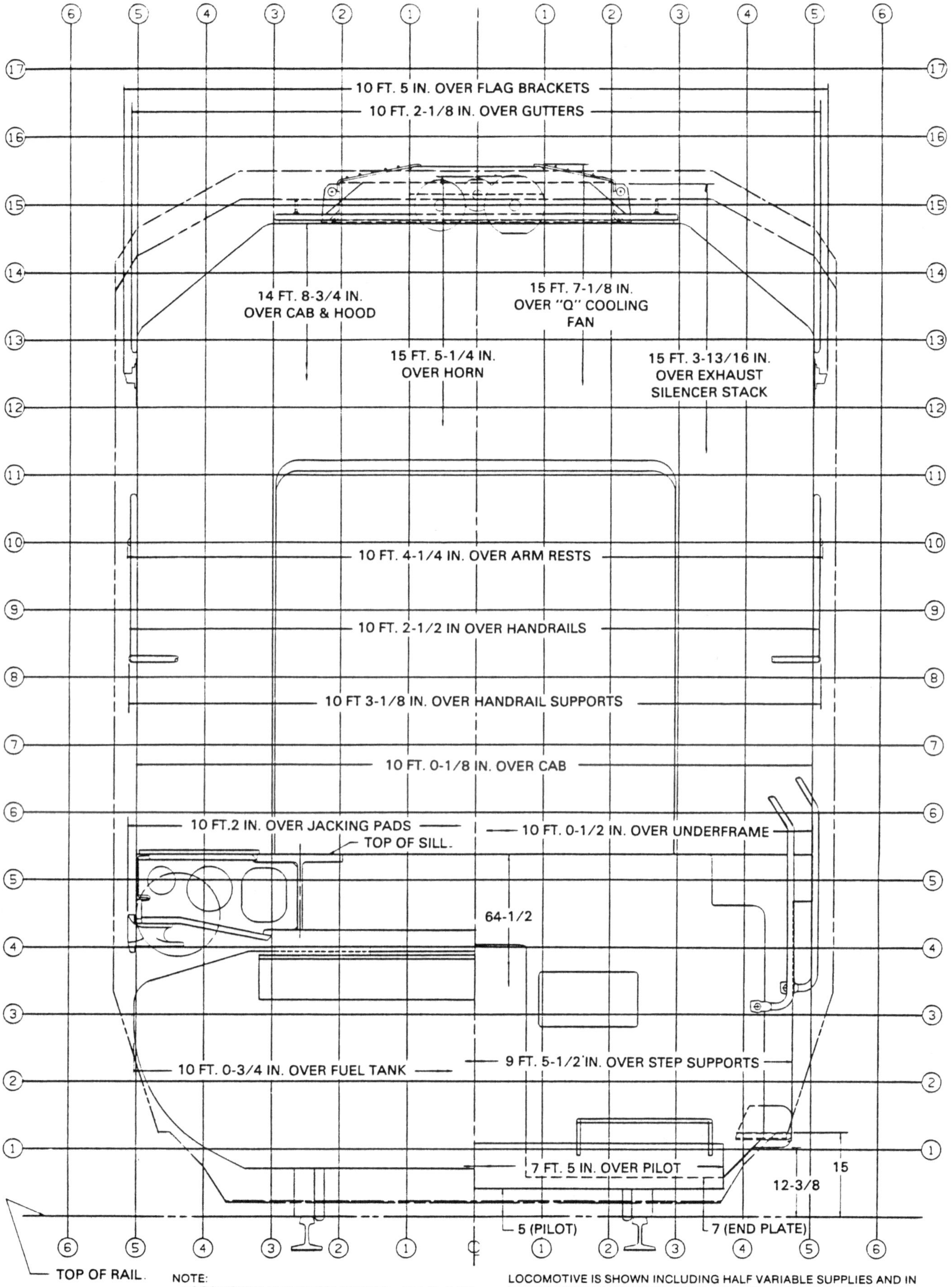

10 FT. 5 IN. OVER FLAG BRACKETS
10 FT. 2-1/8 IN. OVER GUTTERS
14 FT. 8-3/4 IN. OVER CAB & HOOD
15 FT. 7-1/8 IN. OVER "Q" COOLING FAN
15 FT. 5-1/4 IN. OVER HORN
15 FT. 3-13/16 IN. OVER EXHAUST SILENCER STACK
10 FT. 4-1/4 IN. OVER ARM RESTS
10 FT. 2-1/2 IN OVER HANDRAILS
10 FT 3-1/8 IN. OVER HANDRAIL SUPPORTS
10 FT. 0-1/8 IN. OVER CAB
10 FT.2 IN. OVER JACKING PADS
10 FT. 0-1/2 IN. OVER UNDERFRAME
TOP OF SILL.
64-1/2
10 FT. 0-3/4 IN. OVER FUEL TANK
9 FT. 5-1/2 IN. OVER STEP SUPPORTS
7 FT. 5 IN. OVER PILOT
15
12-3/8
5 (PILOT)
7 (END PLATE)
TOP OF RAIL.

NOTE:
LOCOMOTIVE HEIGHT TOLERANCE = ±2 1/2 - 1 1/2
LOCOMOTIVE WIDTH TOLERANCE = ±1/2
LATERAL (BOL. = JRLS. WITH NEW TRUCKS) = ±2 1/4
CLEARANCE OUTLINE:
A.A.R. PLATE "B" DATED MARCH 1, 1972
A.A.R. PLATE "C" DATED MARCH 1, 1968

LOCOMOTIVE IS SHOWN INCLUDING HALF VARIABLE SUPPLIES AND IN NEW CONDITION. STANDING STILL ON LEVEL AND TANGENT TRACK. VERTICAL DIMENSIONS WILL BE 1 1/2 IN. LESS WITH FULL WHEEL WEAR. VERTICAL DIMENSIONS CAN ALSO VARY ±1/4 DUE TO VARIABLE SUPPLIES.

General Information and Identification

MODEL SD60 3800 HP Six Motor Diesel-Electric Locomotive.

TYPE Association of American Railroads designation (C-C), Common designation (0660).

ARRANGEMENT The general arrangement of the locomotive is shown on Elevation and Floor Plan Drawing attached.

The locomotive consists of one unit complete with engine, generator, trucks and all necessary accessories for single or multiple unit operation, with a control cab between the long and short hoods.

NOMINAL DIMENSIONS

Distance, pulling face of coupler to centerline of truck	12′ 8″
Distance between bolster centers	45′ 10″
Truck - rigid wheel base	13′ 7-3/8″
Distance, pulling face front coupler to rear coupler	71′ 2″
Width over cab sheeting	10′ 0-1/8″
Width over handrail supports	10′ 3-1/8″
Height, top of rail to top of cooling fan	15′ 7-1/8″
Width over basic arm rests	10′ 4-1/4″

DRIVE

Driving Motors	Six
Driving Wheels	6 Pair
Diameter Wheels	40″

WEIGHTS AND SUPPLIES

Total loaded weight on rails (approx.)	368,000 lbs.
Fuel	3,200 gal.
Sand	56 cu. ft.
Cooling water	250 gal.
Lubricating oil	283 gal.

CLEARANCES Locomotive outline drawing found on page 97 of this book illustrates clearance conditions.

SAFETY APPLIANCES All steps, grab handles and other safety appliances cover EMD interpretation of FRA requirements.

CURVE NEGOTIATION Truck limits single unit curve negotiation to a 29° or 195 ft. radius curve.

Single unit coupled to a 50 ft. car is limited by car coupler to a 16° or 342 ft. radius curve.

Two units coupled in multiple limited by coupler to a 24° or 235 ft. radius curve.

SECTION A Air System

Basic

AIR BRAKES 26L brake schedule including self-lapping independent and standard 26F control valve portions.

BRAKE PIPING Wrought steel pipe with Association of American Railroads fittings are used. Generally, all piping 1/2″ O.D. and under uses nominal size steel tubing with SAE fittings.

CONDUCTOR'S BRAKE VALVE Recessed conductor's brake valve is provided on left side of the cab.

AIR COMPRESSOR One two stage, three cylinder, water cooled direct coupled compressor, having a displacement of 254 cu. ft. per minute at 900 RPM. Compressor is equipped with large oil capacity, full flow lube oil filtration system, gear type oil pump, drilled rods and disposable intake air filter.

Electric air compressor governor adjusted to maintain reservoir pressure between 130 and 140 psi.

GAUGES AND TEST FITTINGS Large 4-1/2″ air gauges fitted with gauge test fittings are standard. Test fitting is also supplied at compressor unloader switch.

MAIN RESERVOIR Two 15″ diameter x 152″ steel reservoirs mounted beneath the underframe. Total capacity: 49,000 cu. in. Number 1 main reservoir equipped with an air operated automatic drain valve. Centrifugal air filters are provided after both main reservoirs.

WARNING DEVICES Three chime diaphragm type air horn, two bells pointing forward and one to the rear with lever operated modulating horn valve. Horn is located on centerline of cab roof.

One 12″ locomotive bell with internal pneumatic ringer, located in underframe.

Optional

- **Engineer alertness systems.**
- **Overspeed limit.**
- **Six-cylinder air compressor.**
- **Air flow indicator.**
- **Automatic drain valves on both main reservoirs.**
- **Timed blowdown of main reservoir filters and drain valves.**
- **Air compressor low oil protection.**
- **A-1 charging cutoff pilot valve for break-in-two protection.**
- **Air compressor synchronization.**
- **Air compressor clutch.**
- **Compatible operation with locomotives equipped with 6BL brake schedule.**

SECTION B Sanding

Basic

CONTROL Sanding system is controlled electrically. Manual sanding to five mph and automatic sanding afterwards as part of "Super Series" are provided. Automatic sanding from brake valve handle emergency position is provided.

SWITCHES Manual directional sanding switch is provided. A separate switch is provided for lead axle sanding only.

SAND TRAPS Eight single line sand traps are provided, four traps for forward movement and four traps for reverse movement. Sand trap cutoff valves are provided. Outside access is provided for trap maintenance.

SAND CAPACITY Two sand boxes with a total capacity of 56 cu. ft.

Sand boxes are filled from the outside of locomotive on top of hoods.

Optional

- **Increased sand box capacity - Total of 36 cu. ft. per end.**
- **Pneumatic trainlined sanding control.**
- **Automatic timed sanding from optional A-1 charging cutoff pilot valve and/or automatic brake valve.**

SECTION C Multiple Unit Control

Basic

EQUIPMENT Multiple control equipment allows operation of two or more units from one cab. Locomotive equipped with one 27 point power plant receptacle per end, and one power plant jumper cable.

END ARRANGEMENT Power plant receptacle is located in end plate at both ends.

A solid (fixed) multiple unit walkway ramp is provided at both ends.

End arrangement includes multiple unit hand railing and guard chains at both ends.

Optional

- **"Breakaway" support posts for end arrangement guard chains.**
- **Permanent 27 point jumper cables with dummy receptacles are available.**

SECTION D Dynamic Brakes OPTIONAL

- Two-speed dynamic brakes use the traction motors as generators, with the power being dissipated through force ventilated grid resistors mounted in hatch behind cab ahead of inertial filter compartment. Variable voltage type control is standard with dynamic brakes. Parallel grid connection to improve wheel slide protection is provided. Ground relay protection and enforced time delay for power to dynamic brake is standard.

  Positive indication of "power" or "dynamic brake" mode of operation is clearly shown at controller. 4-1/2" zero-center ammeter contains both "brake" and "power" indication, reading in opposite directions from center for each function.

- A grid current trainline control feature is available, if desired.

- Extended range dynamic braking providing high brake effort at low speed is available.

- A self load test feature permitting locomotive loading on its own dynamic brake grids is available.

- Grid blower protection is available preventing dynamic brake operation as result of stalled blower motor or no motor current.

SECTION E Electrical System

Basic

MAIN GENERATOR EMD AC main generator, with a split conductor coil and fan blades on the rotor for increased cooling provides rectified output for delivery to traction motors; 600 volt (nominal) direct current rating, ventilated by an external blower. Armature shaft supported by single bearing with direct connection to engine crankshaft through alternator rotor and flexible coupling. Adequate capacity to continuously transmit the rated output of the engine under all conditions for which the locomotive is designed.

Optimum transmission efficiency is achieved with generator transition. The AC main generator stator windings are divided into two separate "Y" wound machines with added generator diodes to permit connection in series or parallel mode. An external transition contactor connects the stator windings in parallel for maximum generator current and locomotive starting performance, reconnecting in series for full voltage, full locomotive speed operation.

GENERATOR EXCITATION Excitation for main generator supplied from the auxiliary alternator through silicon controlled rectifiers.

ALTERNATOR EMD 200 volt, 3 phase, 16 pole alternator, built integral with main generator, to supply AC power for engine cooling fan induction motors, main generator excitation, and inertial separator exhaust fan.

LOCOMOTIVE CONTROL Traction motors are connected in permanent full parallel with no motor transition required.

Three on-board microcomputers are used to accomplish the control of the Logic system, Excitation system and Diagnostic/Display system.

The Logic microcomputer controls engine speed, locomotive direction, power or dynamic brake operating mode and traction motor and generator switching.

The Excitation microcomputer controls "Super Series" wheel creep control and the latest fuel economy features.

The Diagnostic and Display microcomputer provides conventional fault annunciations as well as onboard diagnostics, interactive trouble shooting and locomotive event and fault history storage.

EMD "Super Series" wheel slip control system utilizes a controlled wheel creep concept to provide improved locomotive adhesion capability and reduced sand consumption. A radar unit provides very accurate ground speed data for comparison to the speed of each traction motor so each motor can be allowed to creep at a speed which provides maximum adhesion for that axle.

Microcomputer motor temperature sensing feature uses individual traction motor current and kW levels to provide simulation of the traction motor heat

(continued)

Basic

LOCOMOTIVE CONTROL (con't.) characteristics to the control system. The motors are allowed to operate at their full horsepower rating until they reach a predetermined temperature. The system then automatically reduces horsepower to protect the motors.

Microcomputer controlled automatic ground relay reset is provided.

LOAD CONTROL Load control provided to automatically maintain horsepower output in accordance with the published tractive effort characteristics of the locomotive.

ELECTRICAL CONTROL CABINET A totally enclosed, readily accessible cabinet houses the locomotive high and low voltage control equipment. Fault annunciation is provided thru the microcomputer display screen to identify equipment malfunction. The top portion of cabinet is cooled by air supplied by the traction motor blower and filtered thru a four element filter using standard pleated paper filters. The bottom portion will be cooled by air supplied by the generator blower and filtered thru a single element fiberglass bag filter box.

Control equipment includes three microcomputers, high capacity power contactors and gang operated reverser and transfer switches.

An additional cabinet, mounted in the engine room, houses the control equipment for the radiator cooling fan motors. Fuse protection is provided for cooling fan motors.

TRACTION MOTORS Six D87B EMD direct current, series wound, force ventilated, axle hung motors with roller type armature bearings. Motors are equipped with increased capacity axle support bearing caps. The D87B motor is equipped with a new four-brush brushholder for improved commutation.

STORAGE BATTERY 32 cell, 64 volt, 466 ampere hour capacity (8 hour rating) unitized battery provided.

BATTERY BOX Two battery boxes are provided, one on each side of the short hood. Trap doors in catwalk provided for servicing and hinged doors provided for removing batteries. Ventilation and drainage provided. Battery boxes are sized to fit either 17 or 25 plate batteries.

AUXILIARY GENERATOR 18 kW alternating current brushless auxiliary generator (with output rectified to direct current), driven from engine gear train provides current for control circuits, lighting and battery charging. Voltage automatically controlled by static voltage regulator module.

Optional

- **Lockout of individual traction motors.**
- **Manual power reduction (hump control).**

SECTION F Engine System

Basic

ENGINE General Motors sixteen cylinder, 2 cycle diesel engine rated at 3950 BHP at 900 RPM with a fuel efficient high capacity gear train turbocharger providing scavenging through the cylinder wall intake ports and multi-valve exhaust.

Power assemblies arranged in 45° "V," with 9-1/16″ bore and 11″ stroke, are secured with one piece crab plates and high strength necked down crab bolts. 9/16″ diameter plunger unit fuel injectors, crowned rocking arm rollers and spall resistant cam shafts are provided. Cylinder liners have laser hardened port and upper bore areas. Cylinder heads are water cooled with thin decks. Pistons are oil cooled. Connecting rods are drop forged. Piston is of the floating design. Piston pins are of the rocking pin design. Heavy base rail Crankcase has wide foot "A" frames. Gear type damper operates on engine oil. Isochronous governor speed control, low engine idle speed, separate overspeed trip and high crankcase pressure protection are included. Engine is equipped with a bolted-on crankshaft stubshaft at the accessory end to prevent engine damage due to accessory failure.

CARBODY FILTERS An inertial separator, roof mounted in a separate compartment behind the cab, supplies filtered intake air to major components. The elements are removed through a hatch in the top of the locomotive. The separated contaminants are blown out by an alternating current fan incorporated in the separator. Filtered air is supplied to the combination traction motor and main generator blower and the engine air filters. Traction motor air is delivered to a duct and plenum chamber system on the underframe and supplies the traction motors with cooling air. The main supply air duct forms the left side walkway.

Generator discharge air is used to pressurize the engine compartment. Traction motor/generator blower is equipped with automatic adjustable air flow inlet guide vanes for improved fuel efficiency.

ENGINE AIR INTAKE FILTERS Disposable paper or fiberglass filter elements with microprocessor display restriction indication provided for engine intake air.

ENGINE FUEL SYSTEM Return flow, single direct current motor driven gear pump, protected by suction strainer and dual primary discharge filters with filter by-pass and indicator to insure clean fuel for the engine. Sight glasses permit visual inspection of fuel flow, and relief valve offers protection against excessive pressures.

ENGINE LUBRICATION The engine lubricating oil system is a pressure system using two positive displacement gear type pumps combined in a single unit. One pump delivers oil for the pressure lubricating system, the other for piston cooling. The oil supply to these pumps is drawn from the oil strainer chamber through a common suction pipe.

A scavenging oil pump is used to draw oil from the engine oil pan through a strainer, pump it through the improved ten element top fill full flow lube oil

(continued)

Basic

ENGINE LUBRICATION (con't.) filter to the round tube mechanical cooler core section of the cooler tank and return it to the strainer chamber. Low oil pressure and high oil temperature protection are provided, resulting in engine shutdown.

The top fill lube oil filter and new positive sealing valves in the strainer box prevent "leak-down" of lube oil while the engine is shut down and ensures that engine lube oil demands are quickly met during start-up.

TURBOCHARGER LUBRICATION An engine driven positive displacement gear type pump supplies oil to the turbocharger through secondary filtration. A separate electrically driven cool down pump supplies oil to lubricate the turbine for a definite time period before starting and after stopping engine.

ENGINE COOLING Pressurized cooling system includes two direct driven centrifugal water pumps, six full length and two half length 6-row radiator sections with water flow divided between two banks. Three 48" diameter AC motor driven "Q" type unlatched cooling fans for reduced cooling system noise emission. Cooling fans are of two speed design. Fans located above radiator cores in long hood cooling hatch.

Water tank and water cooled oil cooler mounted on equipment rack. Hot engine alarm, reduction to sixth throttle level from hot engine condition, and low water shutdown protection device provided. Water fill system to prevent inadvertent opening of pressure cap.

Automatic water temperature control provided by cycling side inlet shutters and three unlatched two-speed AC cooling fans.

ENGINE EXHAUST AND SILENCER Four series connected manifolds discharge into turbine of turbocharger which has single exhaust to silencer mounted directly on turbocharger. A turbocharger screen inspection port is provided.

ENGINE STARTING Engine is started using two 32 volt series connected motors, energized by the locomotive storage battery. Engine start switch located at the governor end of the engine.

Optional

- **Engine purge control.**
- **Increased capacity oil pan.**
- **Fuel oil preheater.**
- **Engine turning jack.**
- **Immersion heater.**
- **Starter motor thermal overload protection.**
- **Mechanical bonded radiators.**
- **Inertial blower motor failure protection.**

SECTION G Trucks

Basic

TRUCK ASSEMBLIES Two flexible three motor, six wheel, high traction truck assemblies are provided per locomotive.

Improved journal springs offer a relatively soft suspension with lower wheel load variation to reduce wheel slip. Accessible shock absorbers between truck frame and middle axle journal bearings minimize undesirable vertical bounce. Lateral and vertical snubbing of the carbody is transferred through a larger diameter center bearing and "H" shaped bolster, and obtained from the damping characteristics of the four secondary rubber sandwich type suspension pads between truck frame and bolster.

Each of the traction motors is oriented in the same direction and each motor rests on a separate transom where a special suspension provides a flexible support, dampening the torque shocks of the motor. This one-way motor positioning contributes to minimal weight transfer between axles and offers improved adhesion characteristics and ease of maintenance.

TRUCK BRAKES Single shoe type brake rigging provided on each wheel, operated by four 9″ x 8″ truck frame mounted brake cylinders, 16″ non-metallic brake shoes. Brake cylinders at one end operate two brake shoes each, with a 5.72:1 lever ratio. The remaining two cylinders operate one brake shoe each, with a 2.86:1 lever ratio.

PEDESTALS Lined with "Nylatron" pedestal liners bolted to frame.

PEDESTAL TIE BARS Fitted and applied at the lower end of the pedestal legs, held in position by bolts.

TRUCK CENTER BEARING RECEPTACLE Truck center bearing receptacle provided with wear plates and dust guard.

SIDE BEARINGS Friction type side bearings.

INTERLOCKS Body and truck interlocks provided, serving as antisluing device in case of derailment.

BOLSTER SPRINGS Rubber sandwich type pads.

BRAKE PINS All pins and bushings hardened and ground.

SECTION G Trucks

Basic

HAND BRAKE Hand brake provided for the locomotive operates on two axles of one truck. Both trucks provided with lever for handbrake connection, making trucks interchangeable.

WHEELS Rolled or cast steel, heat treated, rim quenched, 40″ diameter with 2-1/2″ rim. Wheel treads are finished smooth and concentric. AAR diameter index groove is provided to indicate wheel wear.

AXLES Axles with journals to suit Timken taper roller bearings. Axle material conforms to physical properties of current AAR specifications.

JOURNAL BEARINGS Locomotive is equipped with Timken GG, grease lubricated taper roller bearings.

SLACK ADJUSTERS Pin type slack adjusters.

Optional

- **Clasp brake arrangement.**
- **Provision for wheel truing.**
- **Huck fasteners.**
- **Floating traction motor bellows.**
- **Lateral shock absorbers.**
- **Quick access latch type traction motor inspection covers.**
- **High strength alloy steel journal springs.**
- **Axles splined at both ends.**
- **Combination journal adapter for axle driven speed indicator.**

SECTION H Cab

Basic

CONSTRUCTION Cab is of fabricated steel construction. AAR/EMD Phase II Clean Cab features are provided throughout the interior of the cab.

CAB SEATS The two wall mounted upholstered cab seats have forward and backward as well as height adjustments. Both seats can be turned 180°. Arm rests are provided outside the side windows.

SPEED INDICATOR An electronic speed indicator utilizing the radar pickup is mounted on the control stand in front of the engineer.

CAB HEATING AND VENTILATING Two forced air electric cab heaters located in cab front partition and two side wall mounted strip heaters with individual control switches provide a total of 10.1 kW capacity. Fresh air vent is provided.

INSULATION Ceiling is lined with perforated metal for sound, backed by insulation. Acoustic and thermal type insulation is added to the cab side walls and rear partition of the electrical cabinet.

FLOORING The cab floor is comprised of plywood covered with linoleum and is elevated above the top of the underframe. A trap door in cab floor and side drop doors provide access to equipment beneath cab floor.

DOORS Doors are located at diagonally opposite corners leading to platform alongside hoods and are of honeycomb construction. Doors include interior closure handle and rubber hinge guard. Head bump pads are provided over doors.

DOOR LOCKS The cab doors are fitted with an inside latch and provided with a lock.

WINDOWS Divided center window is provided over low short hood. Side windows on both sides of cab are sliding double sash type and fitted with rounded latches. End windows in doors and cab are stationary and set in a rubber retainer. All window glazing meets FRA requirements.

WINDOW WIPERS Total of six air operated window wipers are provided for front and rear windows on both sides of cab and center windshields. Cover is provided over wiper motors and handles are padded. Valves are recessed.

(continued)

SECTION H Cab

Basic

ENGINEER'S CONTROL STATION Control station, located conveniently to the left of the engineer's seat, includes the engine speed throttle, locomotive reverse lever, automatic and independent brake valve. The lever arrangement is such that the throttle must be in idle before the reverse lever can be removed to isolate the controller. The horn valve, bell valve and independent sander switch are also located in the control stand. Horn valve handle is covered with rubber. Recess in control stand for radio.

ENGINEER'S CONTROL SWITCHES Control and lighting switches located within reach of the engineer, including switches for control and fuel pump, generator field, engine run, gauge lights, headlight "bright" front and rear, headlight "dim" front and rear. Engine stop, number and class light, and isolation switches are located on rear cab wall. Cab heater switches are located on cab heaters, providing individual control.

ENGINEER'S INSTRUMENT PANEL A lighted instrument panel is provided on top of the engineer's controller containing 4-1/2" air brake gauges, wheel slip light, sand light, ground relay light, PCS "open" light, and the traction motor load indicating ammeter. Indicator lights on the control stand are the "push-to-test" type.

MISCELLANEOUS Two coat hooks provided in cab.

One adjustable padded sun visor is provided.

Optional

- **Third cab seat.**
- **Water cooler or refrigerator cooler.**
- **Electronic speed recorder in conjunction with electronic speed indicator.**
- **Awnings/Wind deflectors.**
- **Air conditioning system.**

SECTION I Radio OPTIONAL

- **The choice of radio equipment will depend upon customer requirement. Radio applications vary from minimum provision to complete application.**

  Basic engineer's control stand includes recess for radio.

Basic

FRAMING Underframe is of constant section design and serves as main carrying member for hoods, cab and equipment. Two side sills supported by center sills support catwalk along side of hoods. Draft gear pockets are welded to the built-up platform construction between center sills. The structure is all welded construction.

COLLISION POSTS Increased strength collision posts, integral with front hood, are welded directly to underframe.

FLOORING Floor plates with antiskid surface are welded to underframe on end platforms and along side of hoods.

UNDERFRAME CENTER BEARINGS Welded to body bolster assembly.

SHORT HOOD Short hood with separate toilet compartment with access door, floor drain and outside air ventilator. Continuous weld around floor plates and alcove for water cooler. Entrance from cab to short hood is a stair arrangement with full height opening. A removable hatch is provided on top of the short hood to facilitate toilet or water cooler removal.

LONG HOOD The long hood over the power plant, cooling system, inertial filter and dynamic brake compartments is designed to a minimum width to provide a walkway around the hood.

The power plant hood is bolted to the deck and removable to facilitate complete engine changeout. Hatch portion of the power plant hood is bolted and separately removable as an assembly to remove the complete engine where maintenance facility crane height might not permit complete hood removal.

Hinged roof doors over the engine facilitate individual power assembly changeout without necessity to remove complete hatch.

The cooling compartment is welded to the deck and coupled to the power plant compartment using a slipjoint. The cooling fan hatch is separately removable for radiator core changeout.

The inertial filter and dynamic brake grid compartment is welded to the deck and cab structure. It is coupled to the power plant compartment using a slipjoint. A removable hatch over the inertial filters facilitates overhead removal of the filter elements and removal of the complete generator assembly. A separate removable hatch is provided over the engine turbocharger and exhaust silencer to facilitate their changeout.

(continued)

Basic

LONG HOOD (con't.) A compartment is basically provided in the long hood behind the cab to accommodate Locotrol or other optional equipment.

HOOD DOORS All side doors have suitable outside hinges and latches.

LIFTING EYES Provision is made for lifting eyes on power plant and hatches to facilitate handling with a crane.

Lifting eyes are provided in the end sheets at all four corners.

SIDE STEPS Side switchman's and platform mounting steps, meeting FRA requirements for locomotives used in switching service, are provided at each corner of locomotive.

UNCOUPLING DEVICE Each end of the locomotive is provided with a top operating device arranged to operate from either side of the locomotive.

COUPLERS Type "E", 6-1/4" x 8" shank, 28-1/2" long. Maximum operational swing of coupler is 19° to either side of centerline. Maximum free (manual) swing is 4° from center.

DRAFT GEAR National Castings NC-391 rubber draft gear with alignment control.

HEADLIGHT Twin sealed-beam headlights, front and rear, are equipped with two 200 watt, 30 volt sealed-beam units. Bright and dimmer switch for each light provided in operator's cab. MU headlight control switch provided.

CLASSIFICATION LIGHTS Classification lights built into each corner of front and rear hood.

NUMBER BOXES Four lighted number boxes, two on each end of locomotive, mounted at an angle for both forward and side visibility. Number boxes at front (cab) end are provided with outside access doors. Number panels are of fiberglass and plastic laminate.

(continued)

SECTION J Carbody

Basic

LOCOMOTIVE LIGHTING Additional lights and outlets are as follows:

1. Two recessed cab ceiling lights with individual switches.
2. Two engine room lights.
3. Four stepwell lights.
4. Outlet receptacles: one in engine room, one in cab.
5. Two short hood compartment lights.
6. Two platform lights, one each end.

FIRE EXTINGUISHERS Two 20 lb. Ansul, one located in cab, the other in the engine compartment.

JACKING PADS Combination jacking pad and cable sling is provided near each bolster at side sill.

BALLAST The locomotive is basically designed for balance.

Optional

- **Reinforced nose.**
- **Anti-climber.**
- **Snow plows.**
- **NC390 Draft gear.**
- **Type "F" coupler.**
- **Signal light/roof top beacon.**
- **Toilet.**
- **Low level cooling air intake (tunnel modification).**
- **Ground lights.**
- **Marker and flag brackets.**

SECTION K Fuel Tank

Basic

3,200 gallon capacity, fuel tank built of heavy gauge steel, with baffle plates, located underneath the locomotive body. One filling station on each side. Tank equipped with venting, cleanout plug, and nonremovable water drain.

One dial type top mounted fuel gauge on right side of tank and one direct reading type fill sight glass on each side of tank. Each filling station provided with electric emergency fuel cutoff actuating button. Similar push button located in cab. When operated, engine stops immediately.

Optional

- **4000, 4500 and 5000 gallon fuel tanks.**
- **Automatic fill adapters.**
- **100 gallon retention tank (decreases total fuel capacity by 100 gallons).**

SECTION L Styling and Painting

Basic

OUTSIDE FINISH Acrylic lacquer except for the fuel tank, air reservoirs, trucks and other undercarriage surfaces, which are enamel.

Generally, a Basic styling and painting scheme is a two color or simple three color scheme. Handholds and long hood handrails should be the same color as the primary long hood color. Color lines should not run through door locks, hinges, shutters or other difficult to mask areas. Multicolor or exotic design logos or medallions should be optional premasked reflective or non-reflective material rather than paint. EMD should be given license to adjust paint design locations for manufacturing practicability, visual aesthetics and efficiency.

LONG HOOD AND SHORT HOOD Inside of long hood, short hood and components contained therein painted suede gray enamel. All air, fuel, water and lube oil piping color coded at points of connection.

CAB Inside painted suede gray enamel.

UNDERCARRIAGE Black enamel unless otherwise specified.

TRUCKS AND TANKS Black enamel unless otherwise specified.

Optional

- **Permanent front end identification plates.**
- **Polyurethane paint.**
- **Reflective/Non-reflective markings.**

SECTION M Shipment

Basic

CONSIGNMENT AND ROUTING In accordance with written instructions furnished by the customer.

OPERATING SUPPLIES Locomotive is drained of fuel oil, lube oil and water prior to shipment.

Specification Amendment

Electro-Motive Division Locomotive Specification No. 8128 is amended to incorporate certain remanufactured components when a "Replacement" locomotive is purchased.

Items listed below constitute the maximum number of remanufactured components that may be incorporated in the SD60 replacement locomotive.

TRACTION MOTOR PARTS
- Frame assembly
- Commutator

TRUCK ASSEMBLY PARTS
- (Reused truck components will depend on the truck selected)
- 6-7/8" Axles
- Frame, bolster, safety straps, and pedestal tie bars.
- Brake cylinders
- Brake levers and straps
- Hyatt journal box assemblies with new rollers

Performance Data

Basic

OPTIONAL GEAR RATIOS The choice of gear combinations will depend upon the service contemplated.

GEAR RATIO	70:17	69:18	67:19	66:20
MAXIMUM SPEED	70	76	82	88
MINIMUM CONTINUOUS SPEED (MPH)	9.8 @ 100,000# T.E.	10.5 @ 93,100# T.E.	11.4 @ 85,640# T.E.	12.2 @ 80,150# T.E.
MINIMUM CONTINUOUS SPEED AT FULL HORSEPOWER[1]	12.0 @ 98,250# T.E.	12.9 @ 91,460# T.E.	14.1 @ 84,140# T.E.	15.0 @ 78,740# T.E.

Note 1: Full horsepower is maintained at lower speeds within the thermal capacity of the traction motors. Power is automatically reduced when motor thermal limit is reached.

HORSEPOWER RATING The SD60 locomotive develops 3800 nominal horsepower into the generator for traction at 900 RPM of the engine under the following conditions:

60° F air intake temperature
28.86 inches hg barometer (minimum)
0.845 specific gravity fuel (19,350 Btu per lb.)
.78 engine governor rack setting
60° F fuel temperature

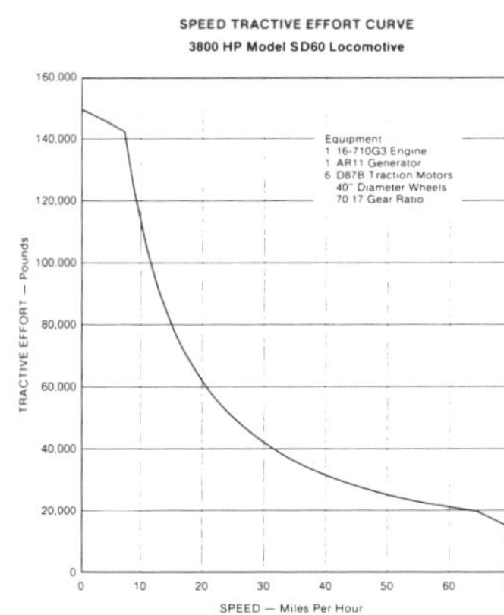

Featuring an on-board computer...

Electro-Motive's 60 Series locomotives take advantage of recent breakthroughs in microprocessor technology. The 60 Series locomotives operate under an advanced computer control system featuring ceramic-based microprocessor chips.

The new computer control system is, in reality, three separate microcomputers, each with its own microprocessor, and each controlling a separate function: logic, excitation, and display/diagnostics. Although each microcomputer performs an independent function, they also do communicate indirectly with each other via a common memory.

Using the multiple microcomputer control system in the 60 Series locomotives has:

- Eliminated 70 percent of the electro-mechanical relays that previously made up much of the logic control network.

60 Series Computer Control System—High Voltage Cabinet Installation

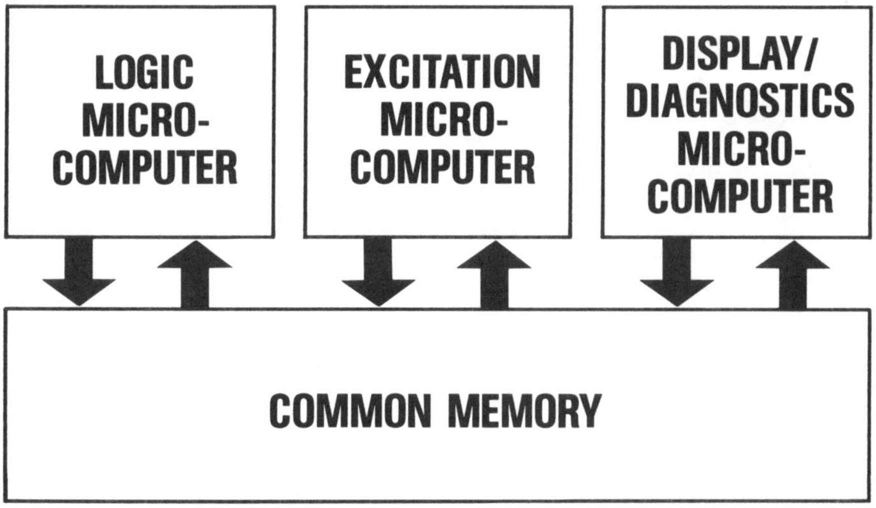

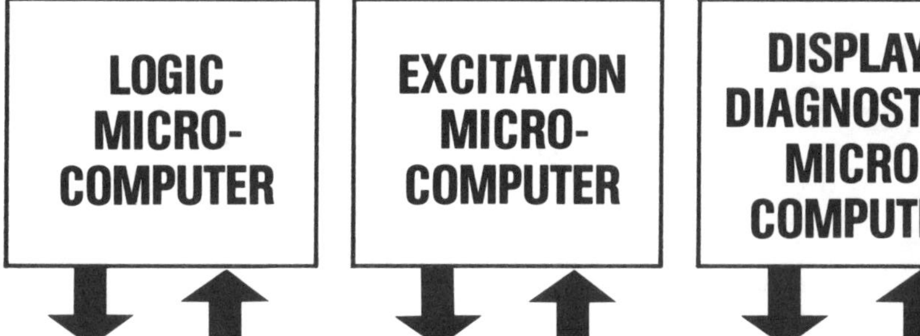

**Block Diagram-
60 Series Computer Control System**

- Reduced electronic components in the power and dynamic brake control system by 21 percent.
- Allowed the introduction of an archival storage feature that simplifies service and repair.

The control functions shared among the three microcomputers are:

- **The LOGIC microcomputer** controls basic locomotive operating parameters, such as engine speed, locomotive direction and traction motor switching.

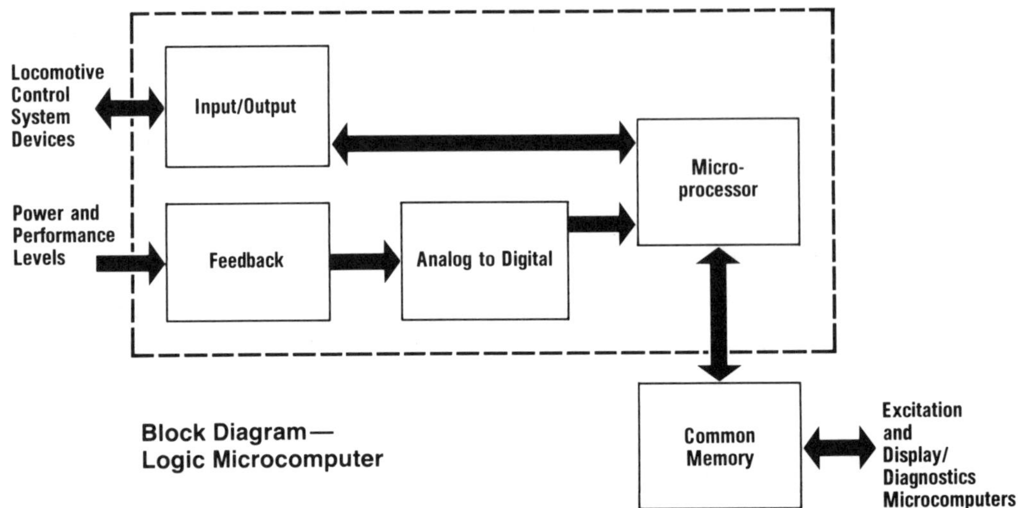

Block Diagram—Logic Microcomputer

- **The EXCITATION microcomputer** controls main generator excitation as required for load control, Super Series adhesion control, and dynamic brake operation.

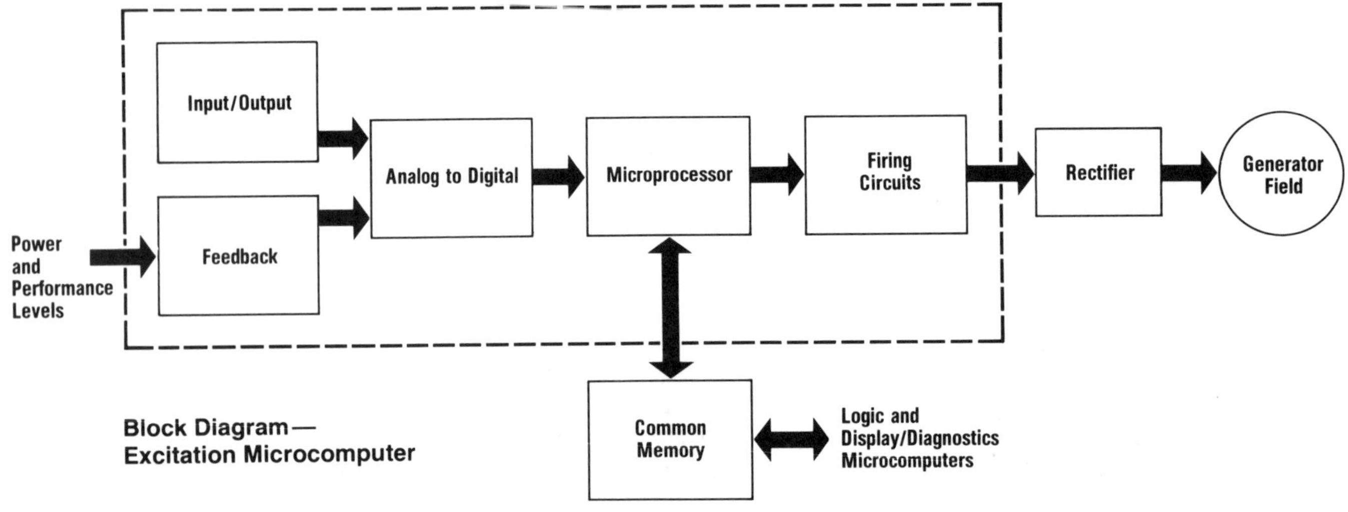

Block Diagram—Excitation Microcomputer

- **The DISPLAY/ DIAGNOSTICS microcomputer** operates the read-out functions for the engineer and maintenance personnel, and records operating data and fault history as a service aid.

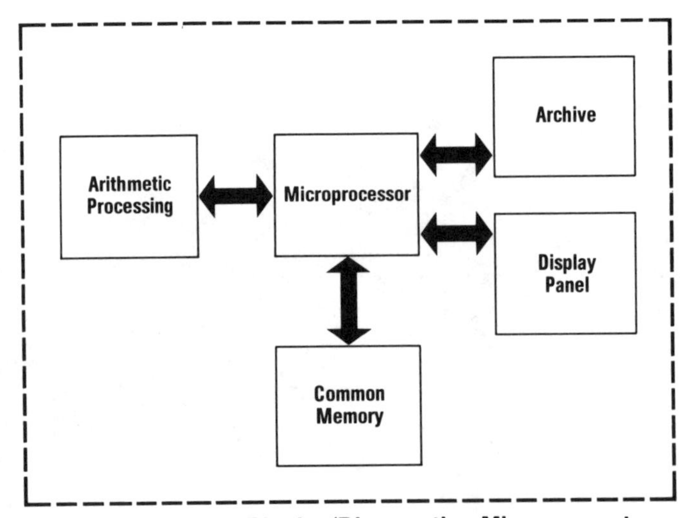

Block Diagram—Display/Diagnostics Microcomputer

It's more than just a computer...

The computer control system does the same jobs as previous Electro-Motive control systems—plus more. Locomotive operating crews don't have to change the way they do things. The only readily visible change will be that operating status and fault messages now appear on the computer display instead of as indicator lights on the engine control panel.

The computer control system functions automatically during locomotive operation:

- Monitoring operator controls as well as various feedback and sensing devices.
- Controlling major locomotive operating systems.
- Detecting and displaying operating status and fault conditions.
- Recording fault messages regarding significant locomotive operating information into archive memory.

The microprocessor computer control system offers many advantages over former locomotive control hardware. The three microcomputers, consisting of 9 standard circuit boards, replace 23 different circuit modules in the earlier analog system, and eliminate the need for most of the 40 to 50 logic relays typically used in mainline freight locomotives.

It performs Logic functions...

The 74-volt locomotive control system devices—including throttle, reverser, and dynamic brake switch—provide inputs to the Logic microcomputer through the Input/Output Modules (IOL). Opto-isolators on the IOL modules convert the signals to 5-volt levels acceptable to the Logic microcomputer Central Processing Unit (CPU). The CPU responds to the engineer's inputs by selecting and operating the proper control mechanisms—including power contactors, switchgear, and engine governor speed solenoids—once again through the IOL modules which, in this case, convert the 5-volt CPU signals to the power levels necessary to operate contactor and solenoid coils.

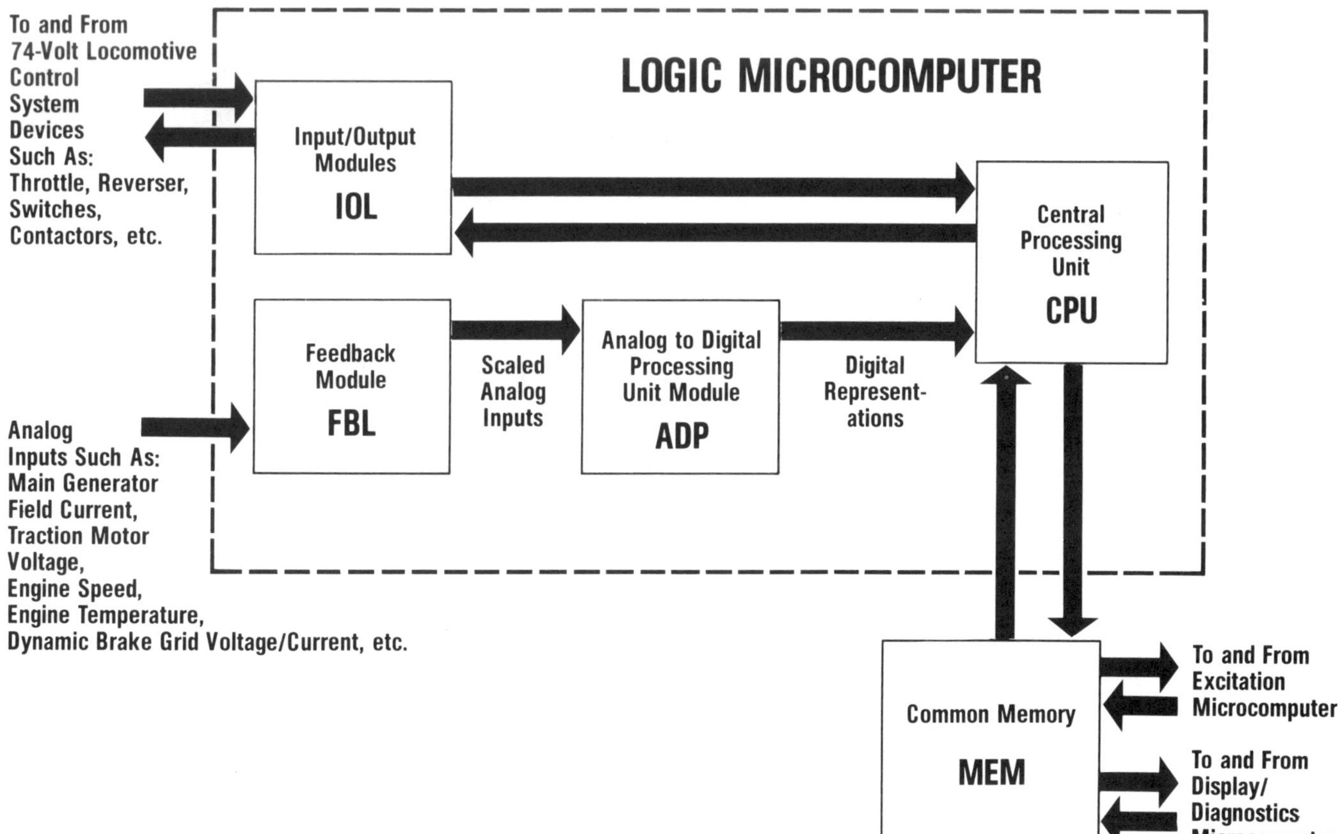

At the same time, the Logic microcomputer tells the Excitation microcomputer, through the Common Memory (MEM), what level of tractive power or dynamic braking the locomotive engineer has selected.

Various analog inputs indicative of significant locomotive operating conditions, such as traction motor voltage, engine speed and engine temperature, are provided through the Feedback Logic Module (FBL). The Analog to Digital Processing Unit Module (ADP) converts these signals to a digital format acceptable by the CPU.

These signals are then also made available, through the Common Memory, to both the Excitation and Display/Diagnostics microcomputers. As the microcomputers cycle through their independent programs and perform their specified tasks, this data is continuously accessible for use.

As an example, engine speed data is used by the Logic microcomputer during engine starting, as part of the "engine purge" function. By controlling starting contactor action, the microcomputer is able to limit engine cranking speed to 30 rpm during the first revolution, thereby preventing internal engine damage due to hydraulic lock.

Engine speed data might also be used by the Display/Diagnostics microcomputer for readout on the display panel during locomotive servicing operations.

Additionally, by monitoring these various locomotive status and operating condition signals, the Logic microcomputer acts as backup protection against catastrophic equipment failures in the event of Excitation microcomputer malfunction.

It controls power and braking levels...

The Excitation microcomputer contains five printed circuit modules. Feedback signals, containing information regarding control handle settings, main generator output, traction motor operation, and dynamic brake grid data, are received either directly through the Feedback module (FBE) or indirectly through the Logic microcomputer and Common Memory (MEM).

When the locomotive is operating in the power mode, the throttle handle position selected by the engineer represents a specific engine speed and output power level. The Logic microcomputer energizes those governor solenoids necessary to set engine speed at the proper rpm.

Simultaneously, via Common Memory, the Excitation microcomputer monitors throttle handle setting, converts the setting to a kilowatt reference point, and sets main generator power output to that level.

The microcomputer controls main generator output by controlling the gating pulses to the silicon controlled

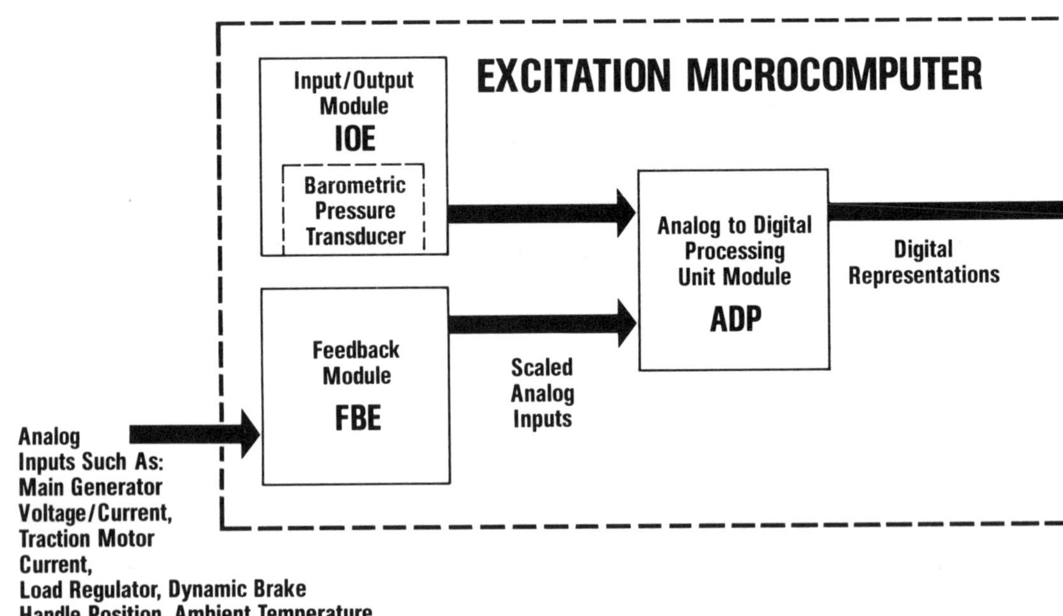

rectifier assembly (SCR), through which main generator field current is supplied.

Using main generator voltage and current feedback signals, the Excitation microcomputer continuously calculates actual generator kilowatt output, compares that value to the established reference point, and adjusts excitation current accordingly.

The Firing Circuit Synchronization module (FCS) is equivalent to the Sensor module on earlier Electro-Motive control systems. The FCS receives feedback pulse trains that are in phase with the positive half cycles of the D18 alternator output voltage. Output pulses, generated by the microprocessor, are phase adjusted to the feedback pulse train via a programmable timer, amplified, isolated, and used as triggering pulses for the SCR.

When the locomotive is in dynamic braking, the Excitation microcomputer converts the brake handle position to a traction motor armature current level reference point, representing a specific braking effort or retarding force. During braking, traction motor armature current level is proportional to the combination of motor field current and motor speed. Motor field current is provided by main generator output.

Therefore, the Excitation microcomputer also controls braking effort by controlling main generator excitation. The microcomputer compares the actual braking effort, as represented by the feedback signal, with the desired reference point and again adjusts excitation current accordingly.

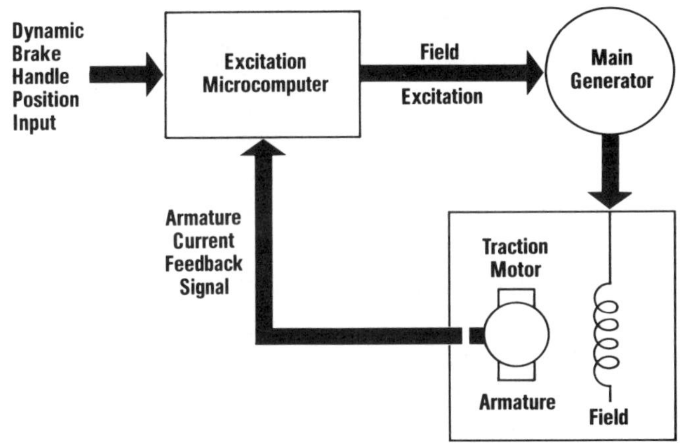

The Excitation microcomputer also controls "wheel creep", which is part of the "Super Series" adhesion control feature. This control system feature prevents the locomotive from "spinning its wheels" by comparing generator voltage to a voltage limit calculated from motor current and radar-sensed ground speed. If a wheel begins slipping, spinning faster than ground speed, traction motor power is automatically limited to a level which allows the wheels to "creep", rather than slip uncontrollably. This provides for locomotive operation at the maximum adhesion level available under existing rail conditions.

The Excitation microcomputer also performs a load control system recalibration function based on input data from a barometric pressure transducer located on the Input/Output module (IOE).

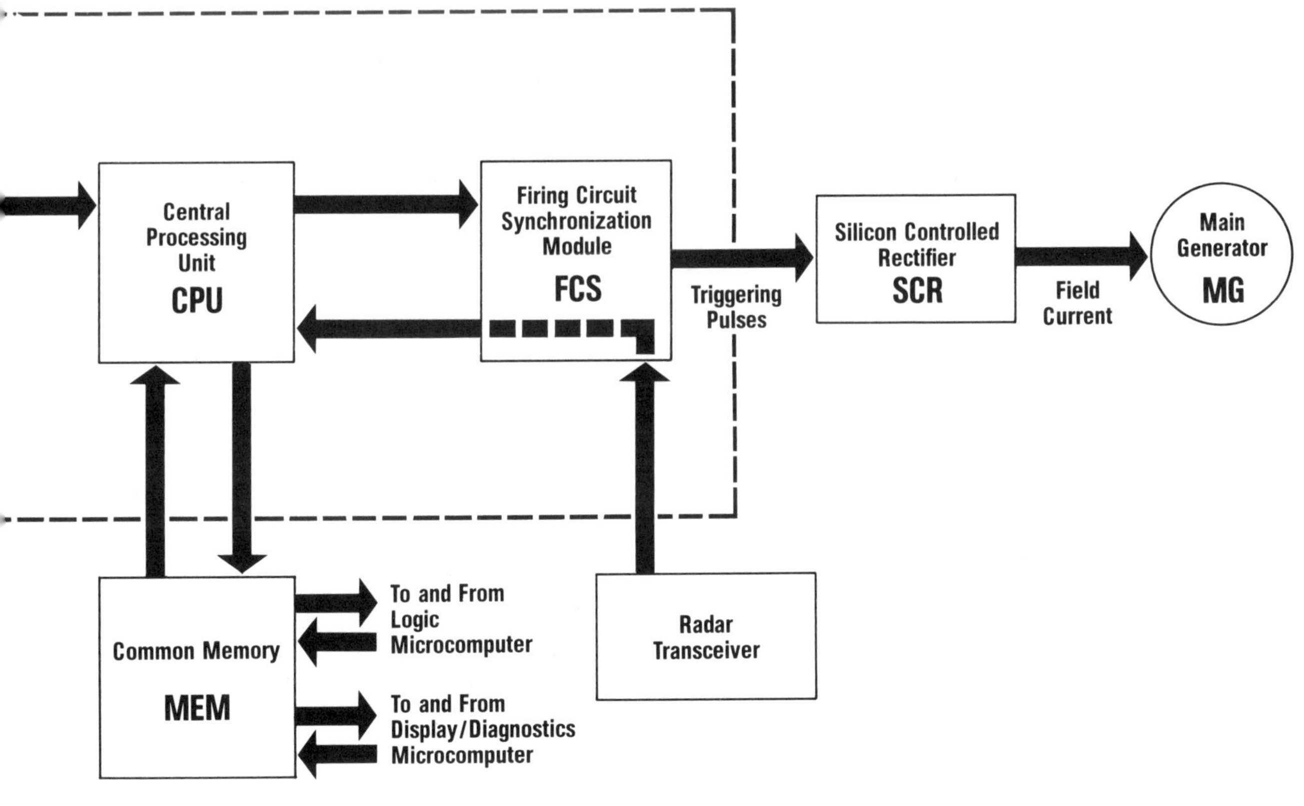

Displays current operating information and provides long-term data storage...

The Display/Diagnostics microcomputer performs the functions which provide the most readily tangible benefits to a locomotive purchaser.

The previously used indicator system consists of 12 to 24 fault indicator lights. This indicator system has been replaced by a single display panel assembly capable of displaying many times more information. Fourteen multifunction control buttons initiate test and maintenance features. All functions are selected through menu-driven software and are displayed on the four-line display panel.

In addition to the display panel, the Display/Diagnostics microcomputer contains four printed circuit modules—a microprocessor (CPU), an Analog to Digital and Arithmetic Processor (ADP), the Archive memory (ARC), and the Common Memory (MEM) used by all three microcomputer systems.

New fault and status indicators can now be added easily since additional software is the primary requirement. Previously, adding just one fault indicator might require a new module, as well as relays and wiring. In addition, the microcomputer can be programmed to analyze fault conditions and provide more detailed information concerning the cause of the problem.

The Display/Diagnostics microcomputer monitors locomotive operation continuously, detects abnormal conditions, initiates corrective action in some cases, and records the condition of the locomotive at the time the fault occurred.

For example, the computer distinguishes between flashover in a traction motor and a true ground—moisture or worn insulation—by the absence of excessive current in the latter case. If flashover is the problem, the ground relay automatically sends a signal to the

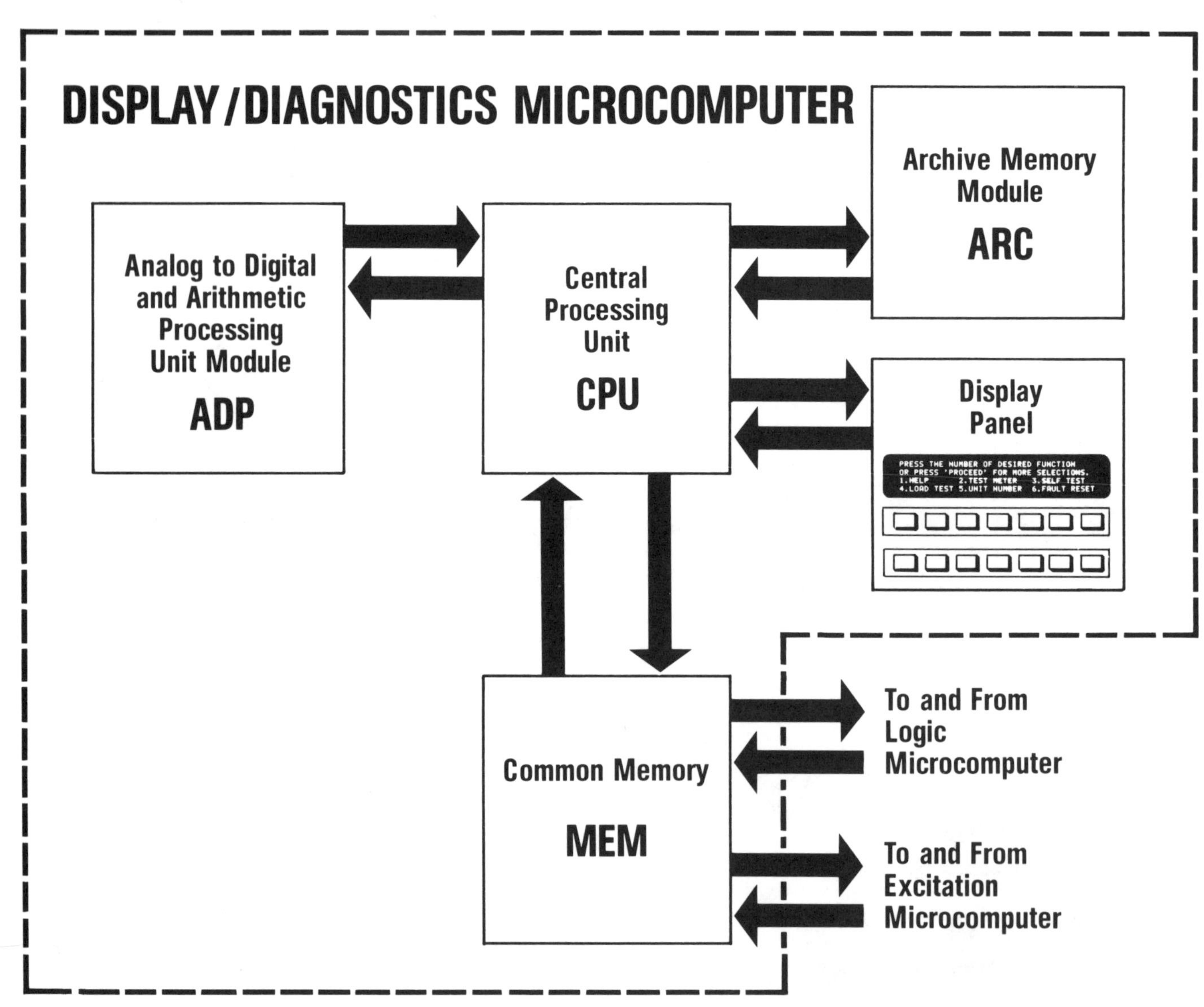

Logic and Excitation microcomputers, which drop field excitation and reduce engine speed. After a ten-second delay, to allow ionized air to dissipate, the microcomputer resets the ground relay and allows the engineer to restore power. After the third fault, however, the microcomputer no longer resets the ground relay. This particular traction motor must be cutout to prevent damage. The Annunciator portion of archive memory records the grounded traction motor fault for display to maintenance personnel.

If the problem is a true ground, the microcomputer reduces voltage. It then tests the circuit by applying progressively lower kilowatt power and allowing a "drying" period.

In this way, the microcomputer distinguishes between moisture (temporary) and worn insulation (permanent). Full voltage is restored when the moisture has dried, or the tests are suspended when continuing negative results indicate permanent insulation damage.

Additionally, electrical components, such as radar, contactors, and the computer system itself, can all be qualified through self-test features of the Display/Diagnostics microcomputer.

Operational running data, such as mileage, kilowatt hours, and duty cycle, are recorded in an archive module. Permanence is assured by battery backup, so that the loss of computer operating power will not affect data stored there. Archive memory is not reset. It always contains the 800 most recent messages, associated data, and the accumulated running data. This data forms an operating history of the locomotive, which can be displayed or transferred, via a communication port, to external equipment. This record of operating data also makes it possible to schedule locomotive maintenance based on actual work performed, rather than on fixed time intervals reflecting average duty cycles, from which individual locomotives often vary widely.

In addition, if a locomotive malfunction is detected by the microcomputer, the time, date, failure mode, and a "snapshot" of critical locomotive parameters are recorded in archive memory. This data is especially helpful when diagnosing intermittent malfunctions.

Becomes a technically advanced maintenance tool...

Service personnel can use the Display/Diagnostics microcomputer to perform a variety of maintenance and troubleshooting tasks:

- Run computer self-test
- Load test the locomotive
- Test several locomotive electrical circuits
- Serve as a multiple test meter, displaying selected signals
- Retrieve data stored in archive memory

The microcomputer is user friendly. It displays conditions in English, not computer codes that need to be deciphered.

During normal road operating conditions, the screen on the Display Panel, located in the center door of the high voltage electrical cabinet, is blank. With previous control systems, if a malfunction was detected and the information was useful to the engineer, an indicator lamp on the engine control panel would be lit. Now a detailed message will appear on the display screen.

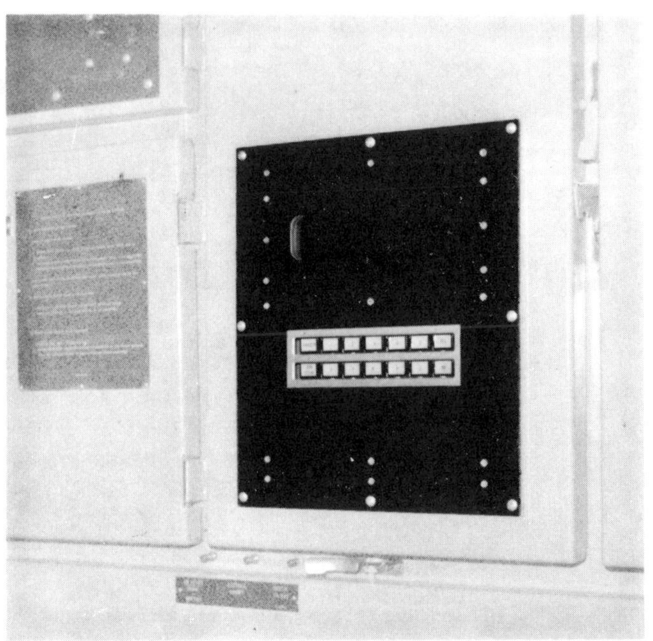

The display panel is located in the upper middle door of the high voltage cabinet—easily viewed by locomotive operating personnel.

The display screen is a vacuum fluorescent dot matrix tube protected with safety glass. The screen can display four lines of 40 alpha-numeric characters per line—a total of 160 characters.

The 14 pushbutton keyboard is utilized during locomotive maintenance. The pushbuttons are labeled with the numbers 0 through 9, PROCEED, RUN/CLEAR, YES and NO.

Typical malfunction and operating status messages include the following:

GROUND RELAY POWER
ENG AIR FILT DIRTY—THROTTLE 6 LIMIT
GOVERNOR SHUTDOWN
ENGINE SPEED INCREASE—LOW WATER TEMP.
LOW COMPRESSOR OIL PRESSURE
EXCITATION LIMIT
TURBO CIRCUIT BREAKER DOWN
GROUND RELAY—TRACTION MOTOR 3

New messages, such as GROUND RELAY—TRACTION MOTOR 3 and ENGINE SPEED INCREASE—LOW WATER TEMP., have been designed to provide improved information to both the operating crew and maintenance personnel. Previous systems merely annunciated that the ground relay had tripped. Today's computer system not only discriminates between power and braking operating modes, but can also, in the case of a flashover, determine the specific traction motor in which the ground fault occurred.

These messages remain on the display screen until the condition has been cleared. In addition, the fault is recorded in archive memory for retrieval at a later date.

During road operation, none of the pushbuttons on the front of the Display Panel are lighted, indicating that they are inoperative. Only lighted pushbuttons are recognized by the microcomputer. The keyboard is enabled through a switch located at the back of the panel. This "enable" switch is accessible only through the relay compartment of the High Voltage Cabinet. Therefore, during locomotive operation, the keyboard pushbuttons are not available for use by the operating crew. The only function the display panel will offer them is displaying operating status and malfunction messages.

Interacts with personnel to perform computer-aided troubleshooting...

During maintenance operations, the service personnel can use the Display/Diagnostics microcomputer to monitor locomotive operation, examine failure history, and test various components and systems. These functions are selected through a menu-type (selection list) system using the fourteen multifunction control pushbuttons.

After the display panel enable switch is turned on, the screen will display the first-half of the main menu.

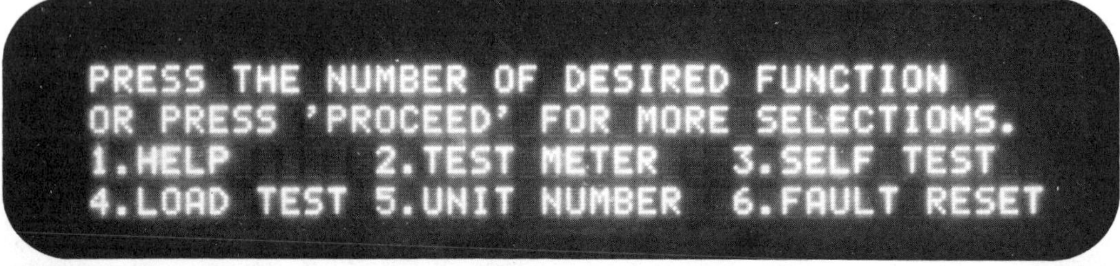

The operator now can select a specific function from this menu. If the first-half of the main menu does not contain the function desired, pressing the PROCEED pushbutton will access the second-half of the main menu.

```
PRESS THE NUMBER OF DESIRED FUNCTION

1.FAULT DATA 2.TIME & TEMP 3.SOFTWARE ID
4.RUN DATA   5.SET TIME    6.TURN OFF
```

MAIN MENU

HELP	Various questions requiring YES or NO inputs...
TEST METER	POWER DYNAMIC BRAKE CREEP CONTROL SELECT METER
SELF TEST	COOLING FANS CONTACTORS RADAR COMPUTER LOAD REGULATOR
LOAD TEST	AUTOMATIC METER DISPLAY MANUAL METER SELECTION
UNIT NUMBER	Display of unit number, capability to change.
FAULT RESET	Display of Fault, capability to reset.
FAULT DATA	ANNUNCIATOR HISTORY TO DISPLAY REPEAT FAILURES HISTORY TO PRINTER
TIME & TEMP	Display of time, date, outside temperature and barometric pressure.
SOFTWARE ID	Display of logic, excitation, and display/diagnostics software program identification codes.
RUNNING DATA	RUNNING TOTALS DUTY CYCLE TRIP MONITOR
SET TIME	Display of computer's internal clock/calendar setting; capability to change.
TURN OFF	Disables computer-aided maintenance menu. Does not affect locomotive control operation.

Main Menu Function Descriptions

The HELP function provides instructions for using the display panel. A series of questions pertaining to the display's functions will appear. The user responds to the questions by pressing either the YES or NO pushbutton. Among the questions displayed are "DO YOU WANT TO DO A SELF-LOAD TEST WITH AUTOMATIC OR MANUAL METER SELECTION?" and "DO YOU WANT TO SEE ANNUNCIATOR FAULTS OR SEE OR PRINT THE FAILURE HISTORY?"

The TEST METER function provides capability of monitoring feedback signals from four different categories. The categories are POWER, DYNAMIC BRAKE, CREEP CONTROL AND SELECT METER.

The feedback signals for POWER are:

THR POS	ID	ENG RPM	200	HRSEPWR	0
MG V	0	MG A	0	KWATTS	0
V REF	0	A REF	0	KW REF	0
TM1 A	0	MPH	0	LR %MAX	0

The feedback signals for DYNAMIC BRAKE are:

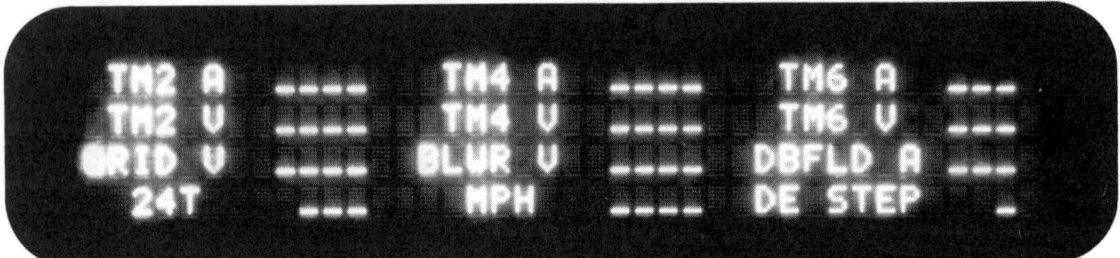

The feedback signals for CREEP CONTROL are:

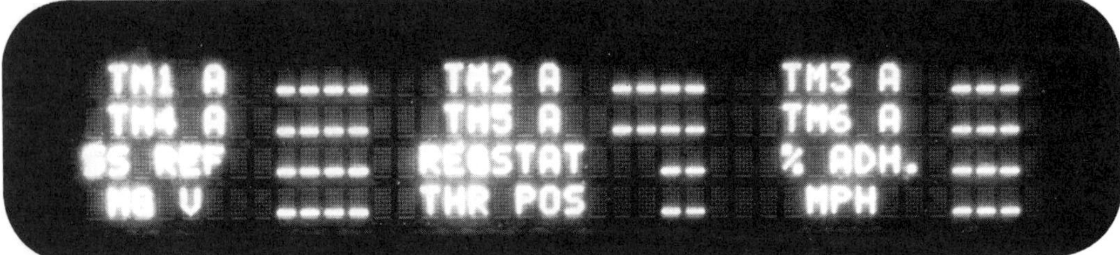

The SELECT METER function allows the operator to assemble a customized set of data for viewing. A maximum of nine signals can be chosen from a selection list of 42 possible choices. The signals available are described in Table I.

The SELF-TEST function permits personnel to automatically test or activate selected circuits or components, such as

```
TEST
1.COOLING FANS     2.CONTACTORS
3.RADAR            4.COMPUTER
5.LOAD REGULATOR
```

These SELF-TEST functions improve locomotive service reliability by helping prevent the dispatch of a locomotive with a control system malfunction.

The COOLING FANS function provides maintenance personnel a convenient method of visually verifying two-speed cooling fan operation and rotation direction. The test is a three-part preset sequence—

1. All fans run at half speed
2. All fans off
3. All fans run at full speed

The test allows sufficient time between sequences for maintenance personnel to visually inspect all fans for speed and direction.

The CONTACTORS function automatically tests the contactors, reverser transfer switchgear, motor/brake transfer switchgear, and relays which were previously tested manually in "circuit check". If a failure is detected, a detailed failure message will appear on the display screen.

In the RADAR test function, a signal having a specific frequency is generated internal to the radar transceiver. This frequency is converted to a speed signal of approximately 40 mph, which is displayed on the screen. After the test is completed, a message will be displayed indicating either a SUCCESSFUL TEST or a fault detected.

The COMPUTER test is a stand-alone test on the three Central Processing Units (CPU's). The test can detect faults with memories, microprocessors, timers and address and data bus transceivers. After the test is completed, a message will again be displayed indicating either that the CPU SELF-TEST WAS SUCCESSFUL or which of the microcomputers failed self-test.

During the LOAD REGULATOR function, the load regulator is driven first to minimum field and then to maximum field. At the completion of the test, the screen will display a message indicating a successful test or a fault detected, such as: LOAD REGULATOR WINDING IS OPEN.

The LOAD TEST function provides two signal monitoring options:

```
LOAD TEST OPTIONS.

1. AUTOMATIC METER DISPLAY
2. MANUAL METER SELECTION
```

When the AUTOMATIC METER DISPLAY option is selected, the control system will automatically transfer to load test if the other necessary operator inputs are present, such as Reverser handle centered, Isolation switch set to RUN, etc.

A predetermined set of signals will appear on the display panel screen.

```
LOAD TEST DATA                        THR POS  08
LR %MAX   100  ENG RPM   900  HRSEPWR 3800
 MG V   1230    MG A   1900  KWATTS  2325
PUSH '2' FOR LOAD TEST 2
```

If the operator selects MANUAL METER SELECTION, the control system will again automatically transfer to load test (if the necessary operator inputs are present), but then will also permit the operator to choose his own selection of up to nine signals (from the same 42 signals available for the SELECT METER function—Table I) to be displayed during the load test.

While running the locomotive load test, the operator can also use the computer to perform "LOAD TEST 2"—a test to verify proper operation of the governor/load regulator system by "allowing" over-horsepower production.

The UNIT NUMBER function provides a display of the identifying number stored in archive memory.

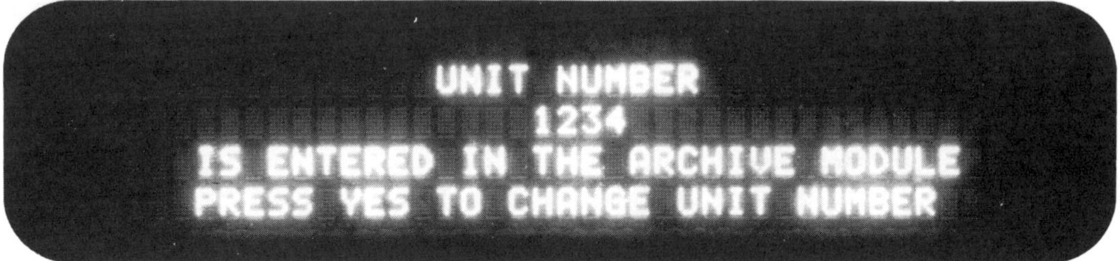

The unit number is required to maintain proper identification of locomotive fault history records. Once an archive module has been installed in a particular locomotive and initialized with its number, that ARC module should, whenever possible, remain installed for the life of that locomotive.

If necessary, the unit number can be changed; however, to prevent an unauthorized change, the operator is also required to use the "enable" pushbutton on the back of the display panel.

The FAULT RESET function allows maintenance personnel to return the computer control system to normal loading after a displayed fault has been corrected.

Certain faults prevent normal locomotive operation. When such a fault occurs, the excitation microcomputer blocks or limits main generator excitation until the fault is actually corrected and the FAULT RESET function used.

A typical example of such a fault is dirty engine air intake filters. When this problem is sensed, the computer system not only displays the message ENG AIR FILT DIRTY—THROTTLE 6 LIMIT to the crew, but also limits locomotive power output. After the filters are replaced and the computer reset using FAULT RESET, the unit will return to normal loading.

The FAULT DATA function provides servicing personnel a selection of four methods to access locomotive fault history—ANNUNCIATOR, HISTORY TO DISPLAY, REPEAT FAILURES, and HISTORY TO PRINTER.

When ANNUNCIATOR is selected, the screen displays a message stating when the annunciator was last reset and the number of faults stored since. By using the PROCEED pushbutton, the operator can view these stored faults in reverse chronological order, beginning with the most recent.

The operator can also obtain more detailed information about a particular fault by pressing the YES pushbutton. The screen will display a "snapshot" data packet containing instantaneous values of feedback signals associated with that fault.

Three types of data packets exist—power, dynamic brake, and engine. The power data packet, shown as an example below, would apply to a LOW HORSEPOWER fault.

After all faults have been accessed, the ANNUNCIATOR can be reset as directed by a message on the display screen. Resetting the annunciator clears it. However, all annunciated faults are still stored in archive memory, accessible by the HISTORY TO DISPLAY function.

The HISTORY TO DISPLAY function allows the operator to view all fault messages stored in archive memory—which has a total capacity of 800 messages and data.

As in the ANNUNCIATOR function, the display sequence is in reverse chronological order, beginning with the most recent, and any applicable data packet can also be viewed when requested.

The REPEAT FAILURES function enables maintenance personnel to choose a particular type of fault and read all messages of that type which are stored in archive memory. A selection list of eight categories of faults is provided: FAILED TO LOAD, IMPROPER LOADING, GROUND RELAY-POWER, GROUND RELAY-DYNAMIC BRAKE, CONTACTOR FAILURES, DYNAMIC BRAKE FAILURES, HOT ENGINE, and ENGINE SHUTDOWN. Once again, the display sequence is reverse order and specific data packets can be viewed when desired.

The HISTORY TO PRINTER function gives servicing personnel the capability of using either a printer to print out, or an external computer to store, all fault messages and associated data stored in archive memory.

A receptacle for connecting the external readout device is provided on the Display/Diagnostics microcomputer CPU module faceplate. Data is read out at a 1200-Baud rate using a 7-bit word length. Messages are "dumped" in reverse chronological order and are "echoed" on the display screen during the transfer.

The TIME & TEMP function provides a one-screen display of current time, date, outside temperature and barometric pressure.

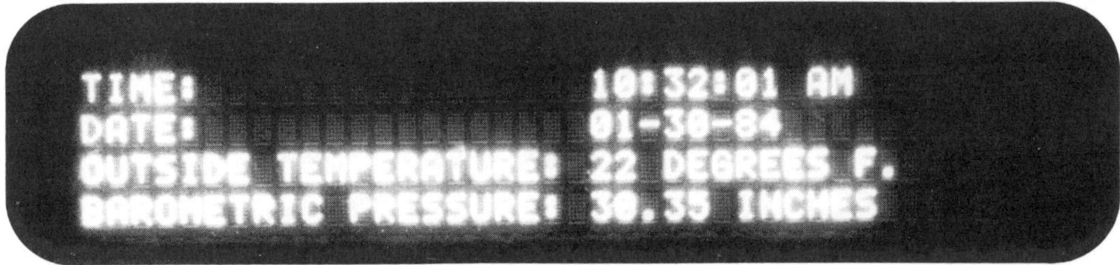

The SOFTWARE ID function is used to determine what specific version of software is presently programmed into the computer system. LCS, ECS, and DCS refer to logic, excitation and display/diagnostics microcomputers respectively.

SYSTEM SOFTWARE IDENTIFICATION
LCS: 08-JAN-85 SK VER.
ECS: 06-JAN-85 PYP VER.
DCS: 09-JAN-85 VHS VER.

The RUNNING DATA function allows maintenance personnel to read specific locomotive running data accumulated in archive memory. This data includes time, distance, power, and throttle position and can be provided in three formats as selected by the service personnel—RUNNING TOTALS, DUTY CYCLE, or TRIP MONITOR. A typical RUNNING TOTALS display is shown below:

ACCUMULATED RUNNING DATA FROM 12-01-83
TO PRESENT. MULTIPLY NUMBERS BY 1000.
KW HOURS 6171 MILES 40.5
HP HOURS 4320 HOURS 2.00

The DUTY CYCLE function is similar to RUNNING TOTALS except that each of the four accumulated totals—kilowatt hours, horsepower hours, miles, and hours—is subdivided into 10 applicable categories: throttle positions 1 through 8, idle, and dynamic brake. Typical display screen is shown:

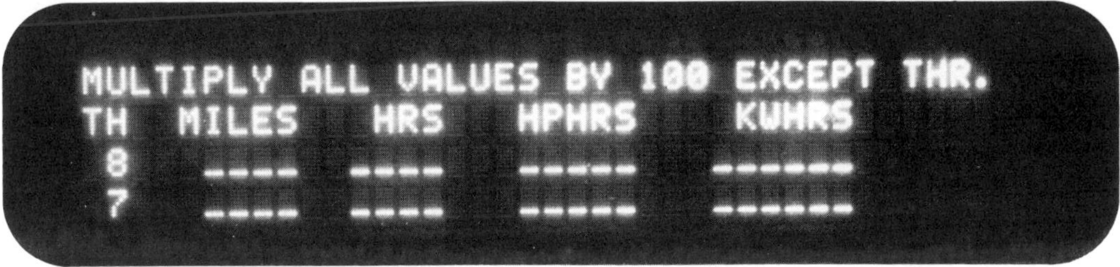

THE TRIP MONITOR FUNCTION provides a data display identical to the DUTY CYCLE function display. However, data is accumulated only for a specific time interval as set up beforehand by the operator. This allows, for example, an ongoing evaluation of duty cycle over one specific run or partial run.

The SET TIME function enables the operator to set or change the computer internal clock/calendar. This determines time and date information subsequently stored in archive memory.

The TURN OFF function stops the computer-aided maintenance and troubleshooting menu program only. It does not affect locomotive operation. Once the menu program is turned off, it can be restarted only through the "enable" switch on the back of the display panel.

TABLE I
Menu of Signals for SELECT METER Function

1. **THR POS**
   Throttle handle position.

2. **ENG RPM**
   Speed of diesel engine; from flywheel magnetic pickup feedback signal.

3. **%ADH**
   Percent adhesion; ratio of tractive effort to weight on drivers.

4. **MPH**
   Locomotive speed; from radar system.

5. **LR %MAX**
   Load regulator position; percentage of maximum field.

6. **HRSEPWR**
   Locomotive horsepower being developed for traction; from main generator output voltage and current.

7. **MG VOLTS**
   Main generator output voltage.

8. **MG AMPS**
   Main generator output current.

9. **KWATTS**
   Main generator power output; from main generator output voltage and current.

10. **VOLTS REF**
    Main generator voltage limit reference established by computer; varies with throttle position.

11. **AMPS REF**
    Main generator current limit reference established by computer; varies with throttle position and load regulator position.

12. **KWATTS REF**
    Main generator current limit reference established by computer; varies with throttle position and load regulator position.

13. **TM1 AMPS**
14. **TM2 AMPS**
15. **TM3 AMPS**
16. **TM4 AMPS**
17. **TM5 AMPS**
18. **TM6 AMPS**

    (13–18) Current through traction motor 1 through 6, respectively.

19. **24T**
    Dynamic brake handle position.

20. **GRID V**
    Voltage across the major portion of the dynamic brake grids.

21. **BLWR V**
    Voltage across dynamic brake grid blower motor.

22. **TM2 VOLTS**
23. **TM4 VOLTS**
24. **TM6 VOLTS**

    (22–24) Voltage across traction motors 2, 4, and 6, respectively.

25. **GEN FLD A**
Main generator field current.

26. **DB FLD A**
Main generator output current through traction motor field during dynamic braking.

27. **MS TEMP**
Motor simulator temperature calculated from traction motor current, voltage, cooling air temperature and flow, and elapsed time information. Cold traction motors can tolerate more power, so computer allows increased generator output power for low speed drag conditions.

28. **GRID1 A**

29. **GRID2 A**

30. **GRID3 A**

(28–30) Current through dynamic brake grid circuits 1, 2, and 3, respectively.

31. **AMB TMP C**
Ambient temperature, in degrees Centigrade.

32. **HVC TMP C**
High voltage cabinet temperature, in degrees Centigrade.

33. **ENGTMP1 C**
Engine coolant temperature, in degrees Centigrade (Probe 1).

34. **D18 VOLTS**
Companion alternator output voltage.

35. **BAR PRESS**
Barometric pressure, in inches of mercury; used for high altitude horsepower limiting.

36. **ENGTMP2 C**
Engine coolant temperature, in degrees Centigrade (Probe 2).

37. **SS REF**
Main generator output voltage limit reference established by computer; used during Super Series—controlled wheel creep operation.

38. **CALC SP**
Calculated speed of fastest driven wheelset, at tread, in MPH.

39. **RECAL SP**
Recalibrated radar speed, in MPH.

40. **MS LIMIT**
Motor temperature simulator limit signal.

41. **REG STAT**
Regulation state; indicates which type of regulation the computer is performing: SS (Super Series), KW (Kilowatt), I (Current), V (Voltage), GX (Generator Field Current), G (Grid), or F (Field).

42. **2V LCSREF**
2-volt reference signals; indicate Feedback modules installed and operating.

710G Diesel Engine Technical Paper

Third Generation
Medium Speed Diesel Engine
The
710G

The Electro-Motive Division of General Motors Corporation has introduced its latest turbocharged two-stroke diesel engine model—the 710G Series. This engine series represents the latest step in a very active fuel economy improvement program at Electro-Motive, providing a 10 percent improvement in full load fuel efficiency over the 645F turbocharged models introduced in 1980.

The Model 710G Series engine will be produced in 8, 12, 16, and 20 cylinder sizes and is intended for rail, industrial, marine, and oil well drilling applications.

The 710G engine has 9.0625 inch (230.2 mm) bore and 11 inch (279.4 mm) stroke, providing increased capacity relative to its predecessor Model 645 Series engines, from which it evolved via a 1-inch longer stroke.

Enhanced expansion of the gases and scavenging of the cylinder from the longer stroke in combination with a new high efficiency turbocharger and new injector have provided the improved fuel economy. This was accomplished with only a moderate increase in mechanical loading of engine structural and power assembly components but a significant decrease in thermal loading of critical power assembly components.

As with previous models, major emphasis has been placed on reliability, interchangeability, and quality of new components from the design phase through the field prototypes, and into production.

Specifications

The Model 710G Series is a two-stroke cycle, uniflow scavenged, open combustion chamber, poppet valve, 45-degree "V" diesel engine with overhead camshafts. The 710 designation refers to the displacement per cylinder in cubic inches. The suffix "G" refers to the crankcase design.

The compression ratio is 16:1. All Model 710G Series engines are turbocharged and aftercooled.

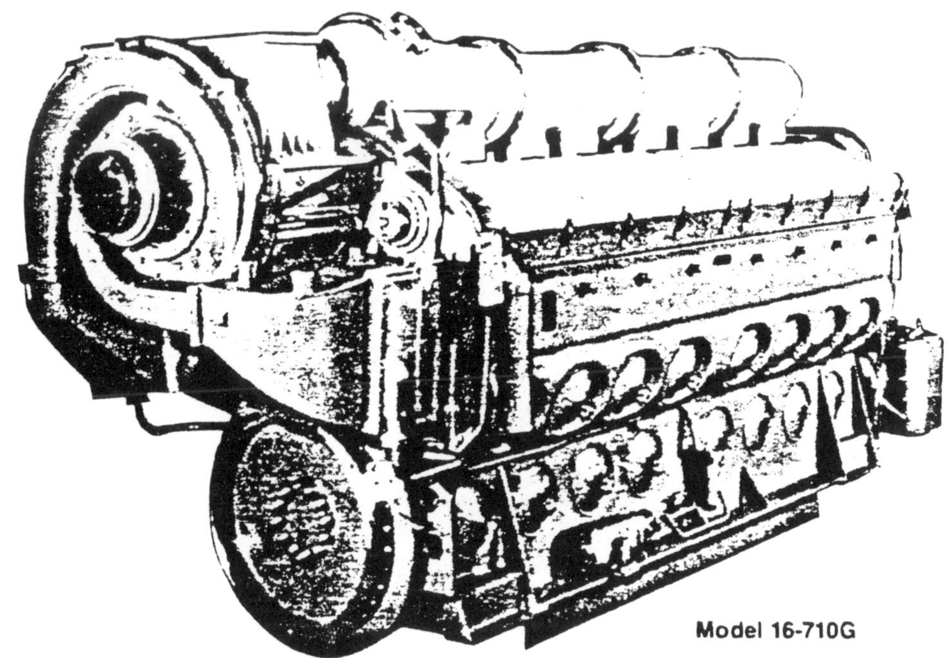

Model 16-710G

16-Cylinder Model 710G Engine Specifications

Bore	9.06 inches (230.2mm)
Stroke	11 inches (279.4mm)
Displacement	710 inches^3 (0.0116m^3)
Cylinder Spacing	16-5/8 inches (422.3mm)
Bank Angle	45 degrees
Compression Ratio	16.0:1
Engine Speed	900 rpm
Brake Horsepower	3950 hp (2945.5kW)
BMEP	153.1 psi (1055.6 KPa)

Model 710G Engine Ratings (Rail Application)

Model	Engine Speed-RPM	Brake Horsepower	Kilowatts	BMEP PSI	BMEP KPa
8-710G	900	2100	1567	162.8	1122.4
12-710G	900	3150	2350	162.8	1122.4
16-710G	900	3950	2947	153.1	1055.6
20-710G	900	4800	3580	148.8	1026.2

Dimensions

The 710G engines are designed to be physically interchangeable with their preceding models. However, in order to accommodate one additional inch of piston stroke and a larger turbocharger, the 710G engines have minor increases in overall dimensions as compared to the 645 Series.

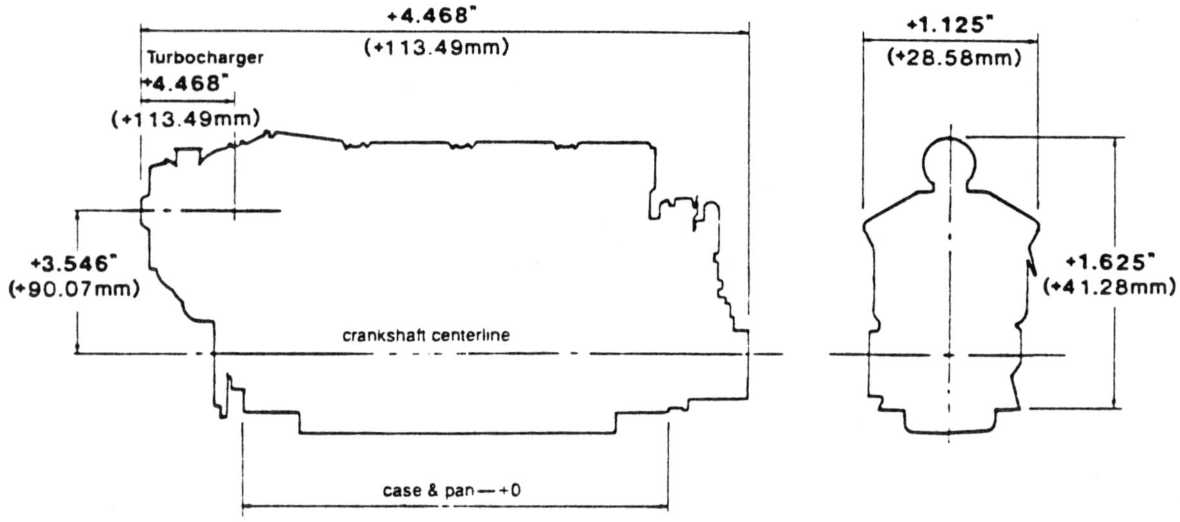

Dimensional increases of 710G engine compared to 645 engine

Horsepower Growth

The Model 567 Roots blown engine was introduced by Electro-Motive in 1938. It was rated 1350 brake horsepower (1007 kW) at 800 rpm. The following chart illustrates the growth in rated horsepower for 16-cylinder rail versions of the 567, 645, and 710G Series engine which have occurred since the Model 567A engine in 1941. Rating increases are indicated for the Model 567B, 567C, 567D3, and finally, for the 567D3A turbocharged engine which was rated 2750 brake horsepower (2052 kW) at 900 rpm. More than 32,500 Model 567 Series engines were produced.

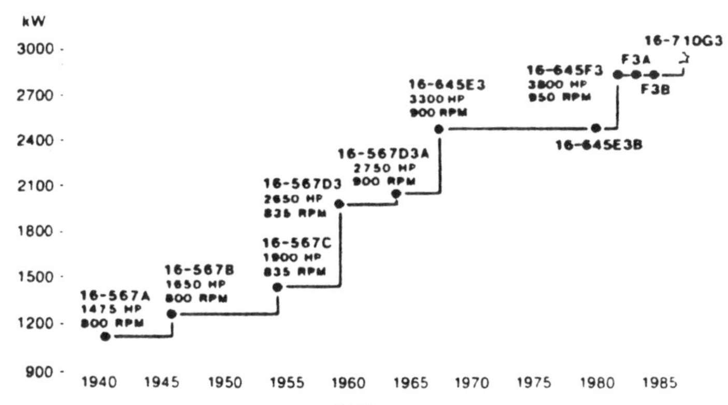

Engine Horsepower Growth Chart

The Model 645E3 Series engines were introduced in 1966, rated 3300 brake horsepower (2462 kW) at 900 rpm. The current production Model 645 Series engine is the 645F3B, rated 3800 brake horsepower (2835 kW) at 950 rpm. More than 26,000 Model 645 Series engines have been produced.

The Model 710G3 engine, rated 3950 brake horsepower (2947 kW) at 900 rpm, represents the highest rating for a 16-cylinder Electro-Motive engine to date.

Performance

The primary goal for developing the 710G engine was to provide improved brake specific fuel consumption. To achieve this goal, many basic performance-related features such as the piston, cylinder liner, injector, turbocharger, and valve train have been evolved from 645 designs for optimum performance. As a result of improved combustion and thermal efficiencies, the 16-710G engine is 10 percent more fuel efficient than the 16-645F model first introduced into production in 1980.

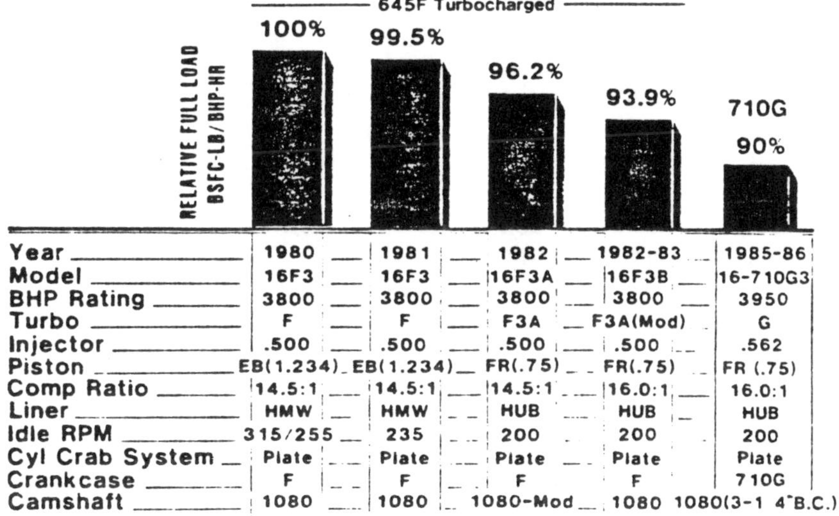

Year	1980	1981	1982	1982-83	1985-86
Model	16F3	16F3	16F3A	16F3B	16-710G3
BHP Rating	3800	3800	3800	3800	3950
Turbo	F	F	F3A	F3A(Mod)	G
Injector	.500	.500	.500	.500	.562
Piston	EB(1.234)	EB(1.234)	FR(.75)	FR(.75)	FR (.75)
Comp Ratio	14.5:1	14.5:1	14.5:1	16.0:1	16.0:1
Liner	HMW	HMW	HUB	HUB	HUB
Idle RPM	315/255	235	200	200	200
Cyl Crab System	Plate	Plate	Plate	Plate	Plate
Crankcase	F	F	F	F	710G
Camshaft	1080	1080	1080-Mod	1080	1080(3-1 4°B.C.)

Relative Improvements In Full Load Brake Specific Fuel Consumption

Combustion Efficiency

The 710G engine provides excellent combustion efficiency with a compact combustion chamber design, established air motion and match, and well-atomized fuel spray. The engine utilizes a 16:1 compression ratio fire ring piston which has the top ring located 0.75 inches (19.05 mm) from the piston crown.

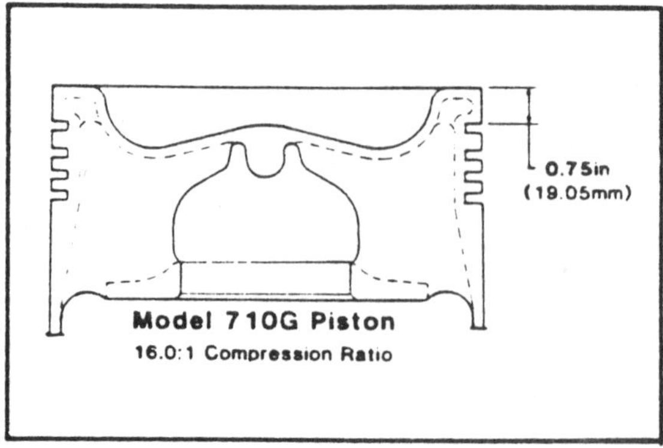

Piston Crown Design

The larger displacement of the 710G engine results in a larger cylinder volume when the intake ports and exhaust valves are closed, which, combined with higher airbox pressures, increases the amount of air retained in the cylinder for combustion.

Control of combustion depends upon successful mixing of air and fuel. Piston bowl shape can have a significant effect on in-cylinder air motion. The 710G piston design takes advantage of a symmetrical combustion chamber consisting of a toroidal shape with high outer rim and a narrow (0.039 in.) piston to head clearance. Scallops or recesses are not needed to accommodate exhaust valves for this two-cycle engine.

The rate of air rotation, or swirl, is readily controlled by the angle of the air inlet ports. As a result of experimental development, the optimum port angle has been established at 15 degrees (measured from radial entry), the same as on the 645 series engine.

Fuel Injection

The 710G engine utilizes the GM unit fuel injector which has been redesigned to incorporate a larger diameter plunger for a higher fuel injection rate. The plunger diameter is 0.5625 inch (14.29 mm) and the injector spray tip contains seven orifices of 0.0152 inch (0.386 mm) diameter drilled at an included angle of 150 degrees. The relatively flat, well-atomized spray produced with this configuration is an excellent match to the combustion chamber over the entire operating speed and load range of the engine. In comparison with the 645FB engine injector, at the same fuel rate, the beginning of injection at full load is retarded approximately 4 degrees and the end of injection is the same for a reduction in injection duration of approximately 4 degrees.

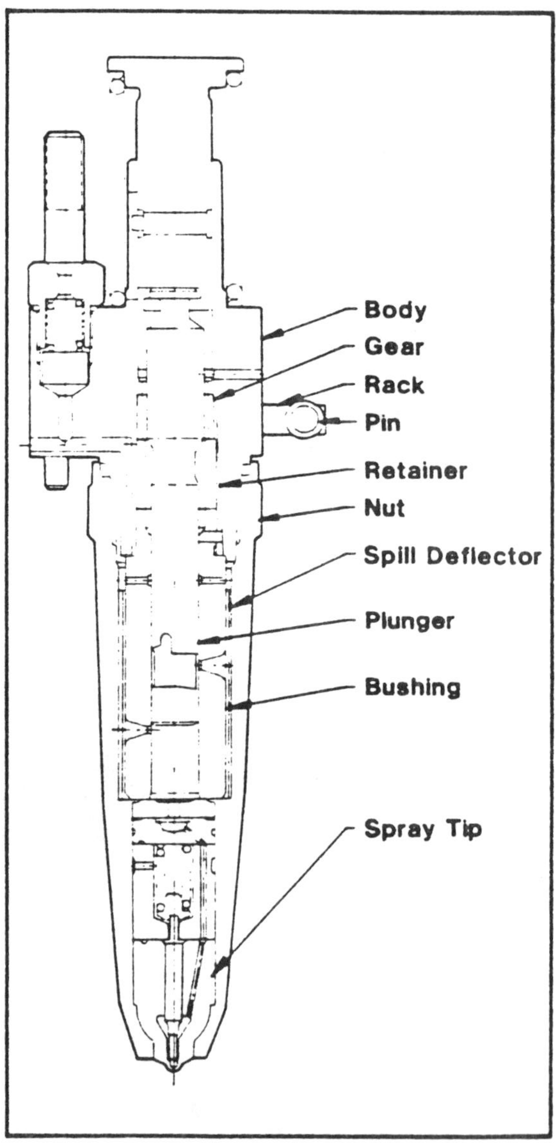

Unit Fuel Injector

Thermal Efficiency

Optimization of the timing events during the engine cycle is of prime importance and directly affects engine performance. In the two stroke cycle uniflow scavenged engine, it is essential that an adequate exhaust lead and blowdown time area be provided to assure that the cylinders are scavenged with a minimum restriction to air and exhaust gas flow and that the cylinders are charged with air with a high level of purity.

Initial performance trends of exhaust valve duration and timing, compression ratio, and liner port timing for the 710G engine were established early through the evaluation of extensive engine test data. Computer modeling of engine performance characteristics further verified these trends and provided useful insight in optimizing the final production configuration of the engine.

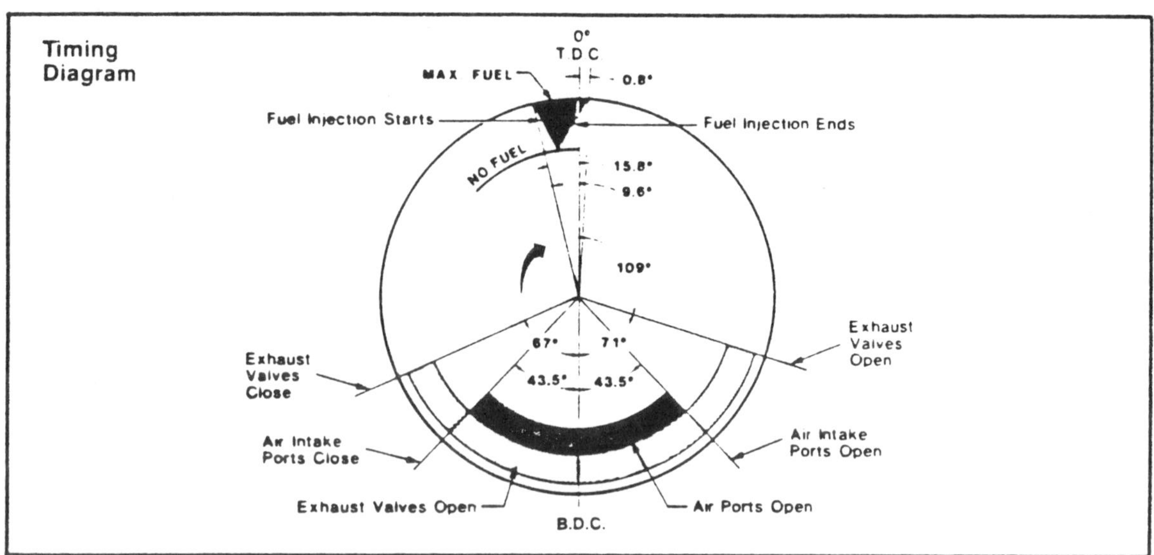

Considerable effort has been expended to minimize the peak pressure increase and thus enhance reliability and durability of critical engine components. The earlier closing of the inlet ports and the later closing of the exhaust valves contribute to the reduction of peak firing pressure. The increased length of stroke coupled with shorter intake port opening duration and later valve timing increases the effective expansion ratio while slightly reducing the effective compression ratio.

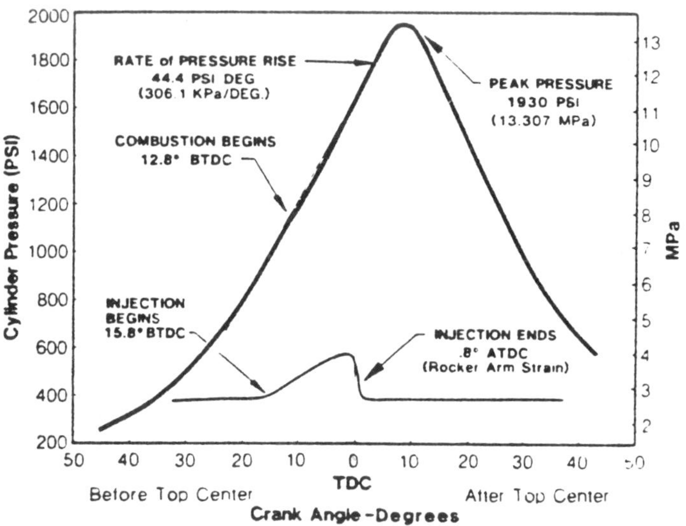

Representative Pressure-Time and Injection Traces

The design of the new injector with optimum beginning of injection at full load, (4 degrees retarded compared to 645) combined with the desirable combustion conditions involving spray atomization and air swirl, assure a satisfactory controlled rate of pressure rise which produces a relatively low maximum pressure and a satisfactory expansion ratio in the cylinder. Representative cylinder pressure time and injection traces indicate a 5 percent increase in peak pressure compared to the 645FB engine.

Fuel Consumption

At rated speed of 900 rpm and rated load of 153 psi BMEP, the 16-cylinder 710G3 rail engine has a specific fuel consumption of 0.323 pounds of fuel per brake horsepower hour (196.4 g/kwh) at 60°F (15.6°C) air in, 29.92 in. Hg (101.04 kPa). The specific fuel consumption values of the 8, 12, and 20 cylinder turbocharged engines are similar.

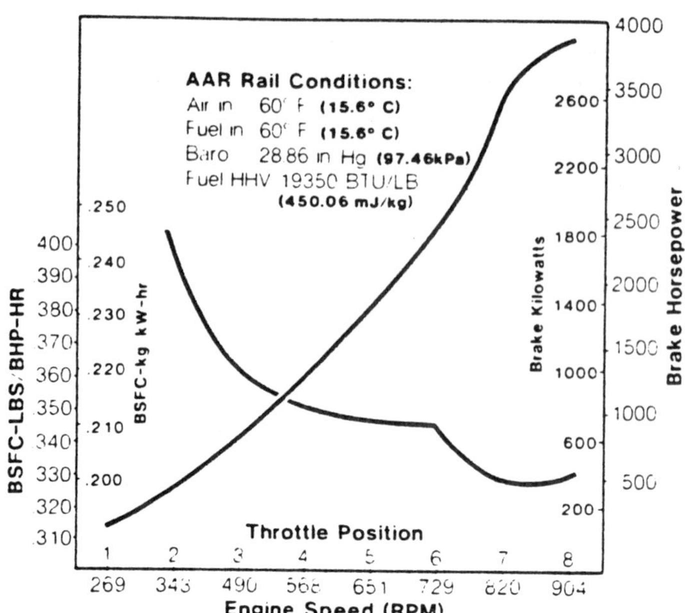

16-Cylinder 710G Rail Engine Brake Horsepower And Fuel Consumption

Exhaust Emissions

The Model 710G engine exhibits lower gaseous emissions than predecessor Model 645 engines.

The emissions rates of carbon monoxide, oxides of nitrogen and hydrocarbons for the 710G engine are 90, 85, and 29 percent, respectively, of the rates for the 645F3B engine.

The low exhaust opacity level of predecessor 645 Series engines is continued in the 710G Series. Opacity at rated output is essentially invisible (four to five percent) at sea level with 90 F (32.2 F) air into the engine.

	710G3 Emissions	UIC/ORE Limit
Carbon Monoxide	0.75	8.0
Oxides of Nitrogen	16.1	20.0
Hydrocarbons	0.14	2.4

Lubricating Oil Performance and Consumption

The 710G engine has demonstrated excellent oil performance. In cyclic durability operation utilizing a 10 TBN (ASTM D-2896), medium viscosity index lubricating oil and averaging 225°F (107°C) oil temperature out of the engine, oil oxidation, as reflected by viscosity increase, amounted to only 10 percent with over 5,000 hours of operation and no oil changes.

Alkalinity levels also remained above recommended condemning limits. Wear metal content in the oil paralleled the 645 engine. Tin, utilized on the 710G piston skirt, also reflected low trace levels in the oil, averaging 10 ppm or less. Oil consumption levels remain comparable to those of the 645 engine at 0.5 percent by volume of fuel consumed.

Based upon 710G engine experience, lubricating oil life is comparable to the 645FB engine — usually more than 9 months between oil changes in rail service.

Turbocharger

The objective of obtaining a large improvement in fuel economy on the 710G Series could not be attained without extensive design and development of the turbocharger compressor and turbine stages, an advancement in the state of the art. The G Series turbochargers provide increased air flow and efficiency improvements without sacrificing the outstanding mechanical features of previous EMD turbochargers.

Beginning with the introduction of the turbocharger in the 567D Series in 1959, EMD turbochargers have employed a positive gear drive through an overrunning clutch. The rotating turbocharger elements are driven mechanically when exhaust heat and engine loads are low to moderate. This important feature is maintained in the G Series design because of its many advantages, especially when compared to the free-wheeling rotors common on four cycle engine turbochargers.

Rapid response from quick air delivery is possible for starting and for suddenly applied load acceptance frequently encountered in many industrial, and oil well drilling applications. In certain emergency stand-by applications the engine must be started, and reach rated engine speed and full load in less than ten seconds.

A gear drive also enables the compressor to provide an adequate air supply to the engine cylinders at part load. Excellent part load fuel economy is thereby obtained.

The G turbocharger had very few design constraints relative to its aerodynamic performance. The 16FA turbocharger was introduced to the 645 Series in 1982. Aerodynamic streamlining was employed to attain a significant four percentage point net gain in turbocharger combined efficiency. The accompanying increase in total air flow improves engine thermal efficiency and fuel economy and contributes to a significant reduction in cylinder exhaust temperature. The resulting reduction in engine component operating temperatures was unprecedented, thereby providing a solid basis for 710G engine component reliability improvements.

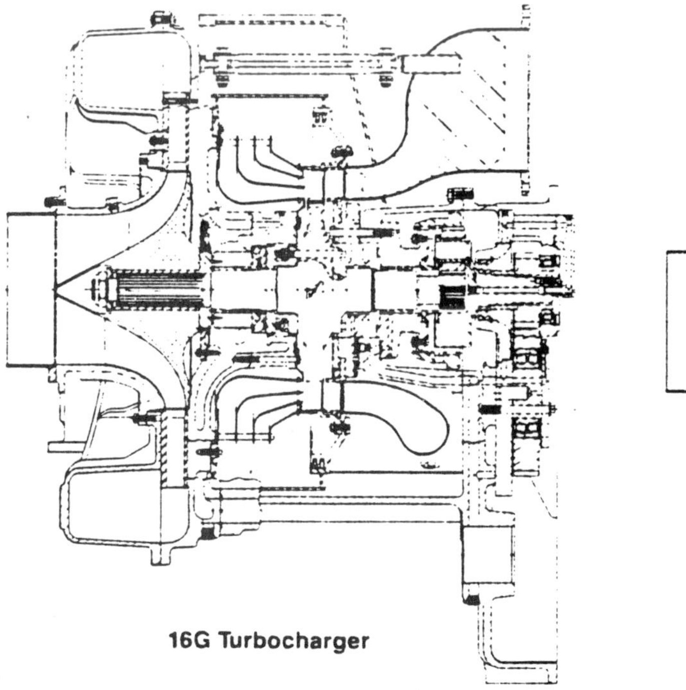

16G Turbocharger

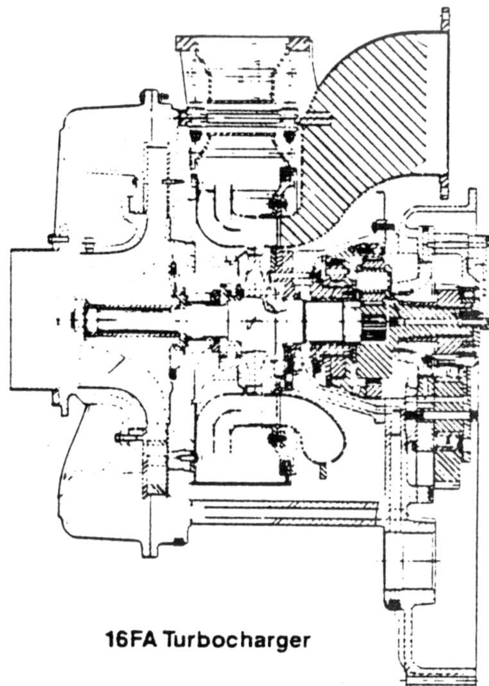

16FA Turbocharger

Compressor

The G Series turbochargers utilize advanced centrifugal compressor impeller/diffuser technology. The compressor stages are under continuous development at EMD enabling an optimization of aerodynamic performance. The compressor performance is a major contributor to fuel economy improvements, and must be attained without sacrificing long term durability and reliability.

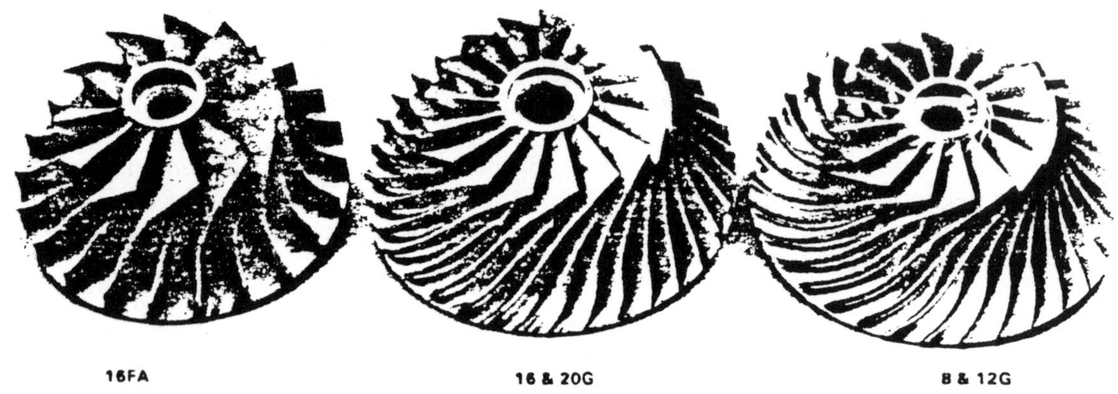

16FA 16 & 20G 8 & 12G

The blading (nozzle vanes and rotor blades) is designed and manufactured to maximize aerodynamic performance. The annular exhaust diffuser and duct assembly utilize split channels for carefully balanced pressure recovery, exhaust gas kinetic energy conversion and improved turbine efficiency.

Since the 16G model has the highest tip speed of any Electro-Motive turbocharger, several mechanical refinements were required. These included a large capacity thrust bearing and integral thrust washer for improved life, increased strength turbine wheel, and optimized three-hook turbine blade/wheel serrations for stress control.

Cylinder Power Assembly

The cylinder power assembly in the 710G engine retains the long proven unitized assembly concept of the EMD engine line. While the assemblies are not directly interchangeable with the 645 engine due to the longer components involved for the 710G, the two assemblies are similar. Unique to the 710G engine are the piston, head, and liner but the remaining components of the power assembly are interchangeable with the 645FB engine.

Piston

Due to the increased displacement of the 710G over the 645, achieved by a 1-inch stroke increase, the piston length has been increased by one inch.

The piston length increase is required to maintain an effective seal between the airbox and oil pan. The inlet ports of the liner must be covered by the piston skirt throughout the engine cycle. Since the stroke of the piston is approximately the distance from the bottom of the air inlet ports to the top of the liner, when piston stroke is increased, the piston length must also be increased by the same amount.

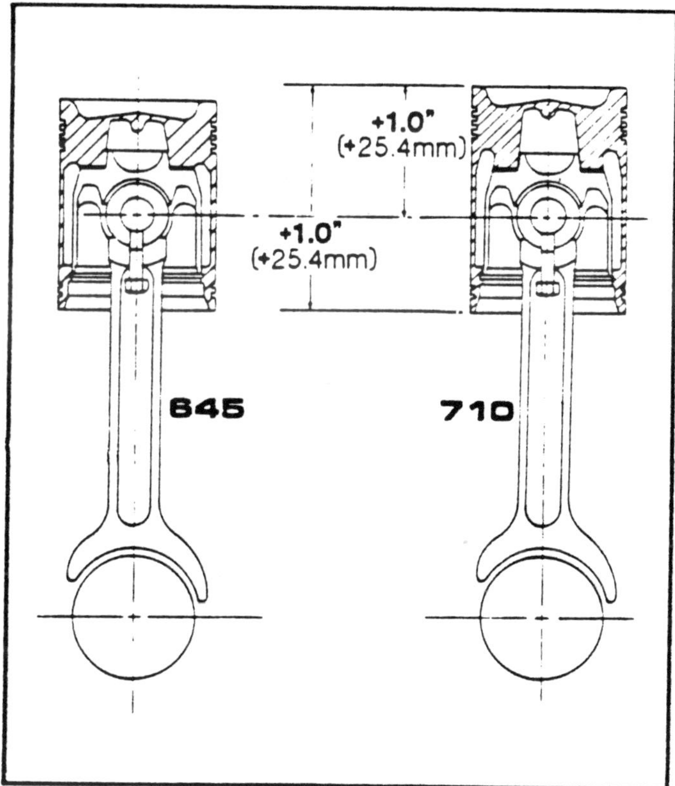

Piston Dimensional Increase

The pin centerline within the piston was lowered by 1 inch with respect to the crown to retain the interchangeability of the pin carrier and connecting rods. The drop of the pin centerline not only provides more room to improve the structural strength under the crown but also improves the piston dynamics and reduces the liner wear.

Compared with the 645 Series, the 710G engine piston has about 20 percent higher side thrust because of higher cylinder pressure and connecting rod angularity. High side thrust increases the potential for liner bore scuffing. To eliminate this possibility, the piston skirt was tin-plated, improving compatibility of piston and liner materials.

Cylinder Liner

As compared to the 645, the 710G cylinder liner has been lengthened by 2 inches. An inch is required from the bottom of the port to top of the liner due to the increased stroke. The additional inch from the bottom of the port to the bottom of the liner is required to contain the longer piston.

With the exception of its increased length, the 710G engine liner retains the features of the 645 liner. The port relief surface and upper bore of the liner are hardened by the laser hardening process pioneered by EMD and currently utilized in all 645 liners for turbocharged engines. Liner bore finishing and all other features of the cylinder liner are analogous to the 645 including the brazed water jacket concept.

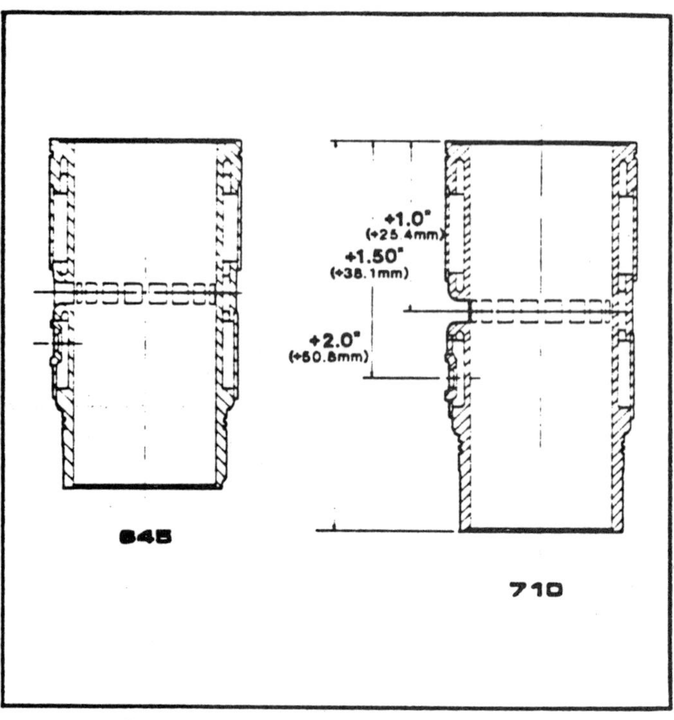

Cylinder Liner Dimensional Increases

Cylinder Head

The 710G cylinder head operating temperature is 30°F lower than that of the 645FB engine. This is due to two factors; first, the increased air flow of the 710G engine Model G turbocharger has reduced the thermal load, and second, the cylinder head has new thickness tolerances made possible by a new machining process. The new machining process has facilitated a reduction in the range of firedeck thickness produced as well as the minimum specified thickness.

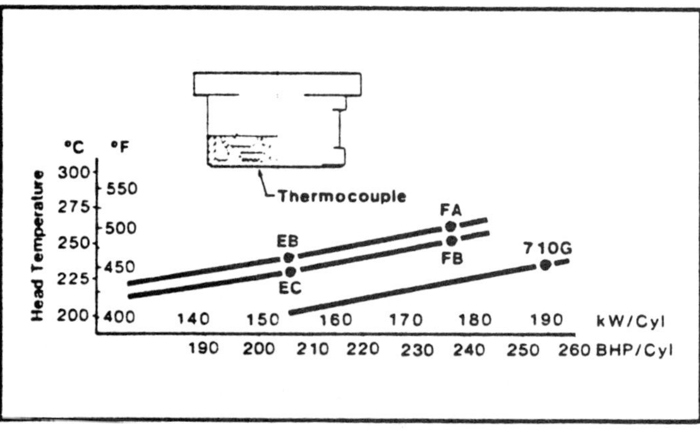

Cylinder Head Thermal Performance

Piston Carrier, Pin and Bearing

The 710G engine retains the "rocking piston pin" long proven in previous EMD engines. The mechanical and thermal loadings of the 710G engine do not represent a significant challenge to the capacities of the rocking pin and bearing.

Structural Features

The 710G engine was designed structurally not only to accommodate the longer stroke but also to provide future capability to withstand even higher peak firing pressures generated during combustion for future growth.

Crankshaft and Main Bearing

The 710G crankshaft incorporates the same 6.5 inch diameter crankpin and a 1 inch larger main bearing journal of 8.5 inches. This geometry allows interchangeability of the 645 fork and blade rod assemblies with the 0.5 inch longer throw of the 710G crankshaft.

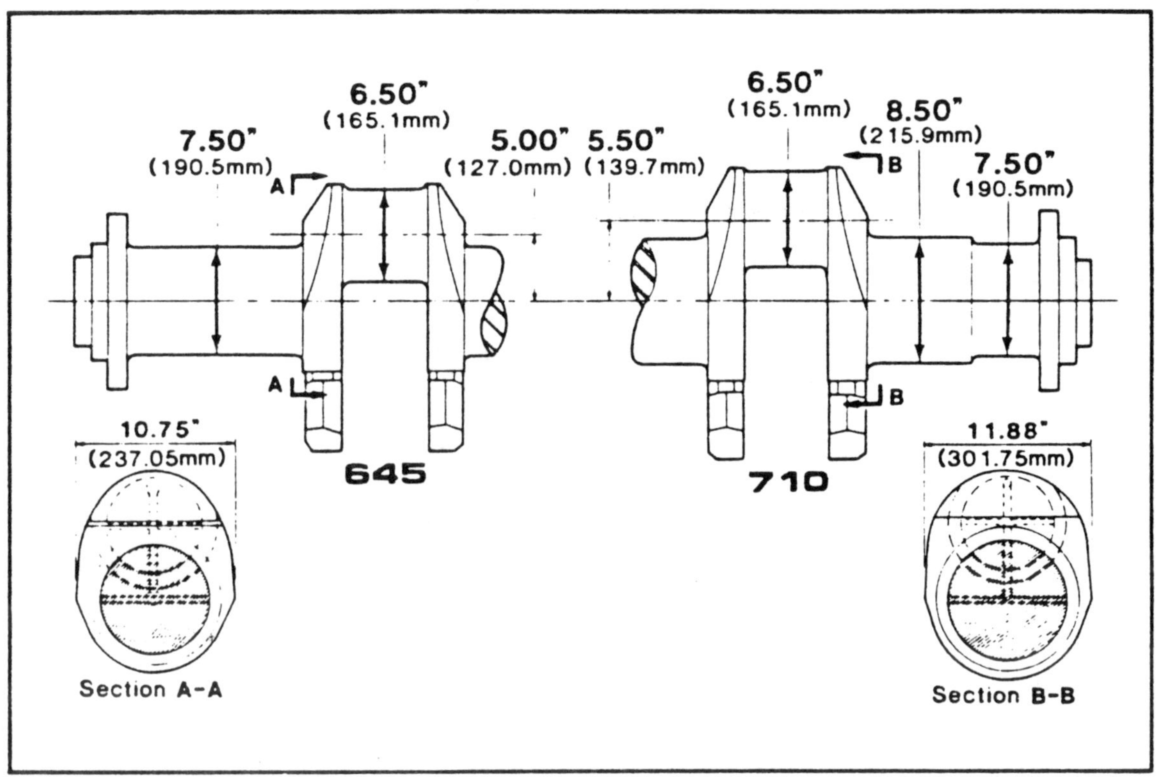

Crankshaft Comparison

The front of the crankshaft has been designed to facilitate the application of higher capacity power-take-off design used on locomotive, marine, and industrial products.

The 710G main bearings are the identical width (axial direction) but 1 inch larger in diameter than their 645 counterparts. These bearings utilize the same high reliability bearing materials as on the 645 design. Bearing geometry (parting line height and diameter clearance) has been defined to utilize the same main bearing torque specifications as 645 engines.

Crankcase and Oil Pan

The 710G crankcase—engine block—is the traditional Electro-Motive fabricated structure. It uses much of the high technology automated welding and machining processes used to manufacture the 645F crankcase. The design features of the 645F crankcase incorporated into the 710G crankcase include:

- Improved mainframe-to-airbox weld attachment and thicker base rail.
- Thicker top deck forging and end bar and heavier weld attachment to end plate.

The 710G crankcase incorporates a new higher strength airbox assembly with 1/16 inch thicker side sheets for higher peak pressure capability. The overall crankcase structure is approximately 1.62 inches taller and 1.12 inches wider than the 645F design. These changes were necessary to accommodate the 1 inch longer stroke and subsequent increases in power assembly component length.

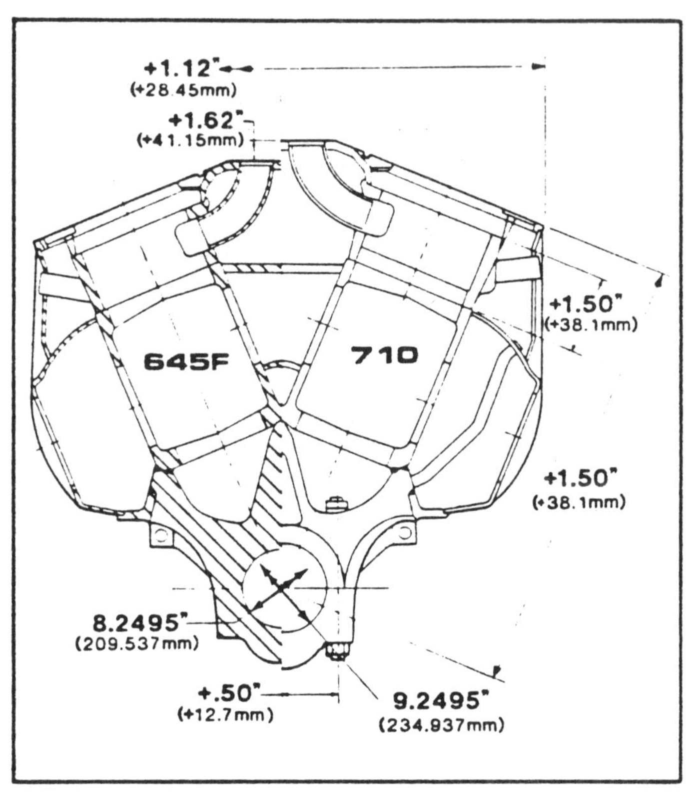

The mainframe forging includes a larger main bearing bore and strengthened main bearing caps. The mainframe weld attachment remains essentially identical to the 645F design.

A new 1.5 inch taller cylinder head retainer forging is used in the 710G crankcase design which includes a significantly improved cylinder head retainer-to-airbox weld attachment. In addition, the taller head retainer requires the use of a 1.5 inch longer power assembly retention (crab) bolt with a larger spherical seat.

Camshaft and Rocker Arm Assembly

The 710G engine uses a 0.5625-inch plunger injector for fuel delivery. Since this injector creates more force in the camshaft system as compared to the 0.5 inch injector, the camshaft system has been redesigned to accommodate the higher loads. Design changes incorporate a 3.25 inch base circle with the previous 645 cam lift profile and 0.375 inch larger diameter roller follower.

The rocker arm assemblies have been strengthened to improve the reliability with this higher loading. The 710G rocker arms utilize the standard 645 rocker arm shaft bushings.

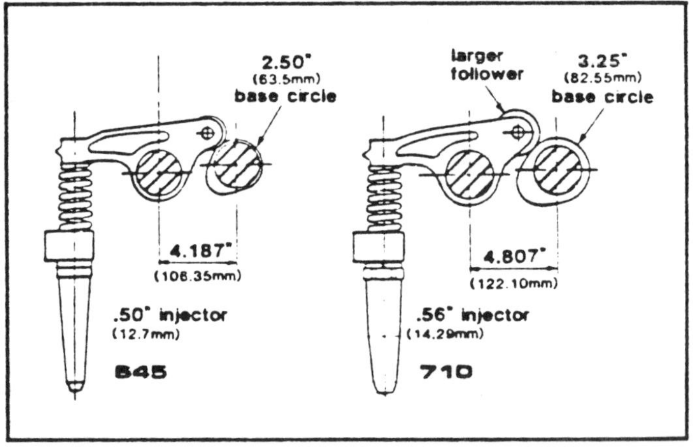

Dimensional changes to Injector, Camshaft and Rocker Arm Assembly

Lube Oil System

The 710G model uses a higher capacity main lube oil delivery system to accommodate the higher oil usage of the 1 inch larger main bearings. Piston cooling oil flow deliveries have also been increased to provide for equivalent delivery of cooling oil flow to the piston carrier with the longer stroke. The increased stroke of the engine and longer power assembly components necessitate a change in the piston cooling tube design. This new design incorporates a higher strength investment casting to support the shorter flow nozzle and provide the bolting attachment to the bottom of the liner.

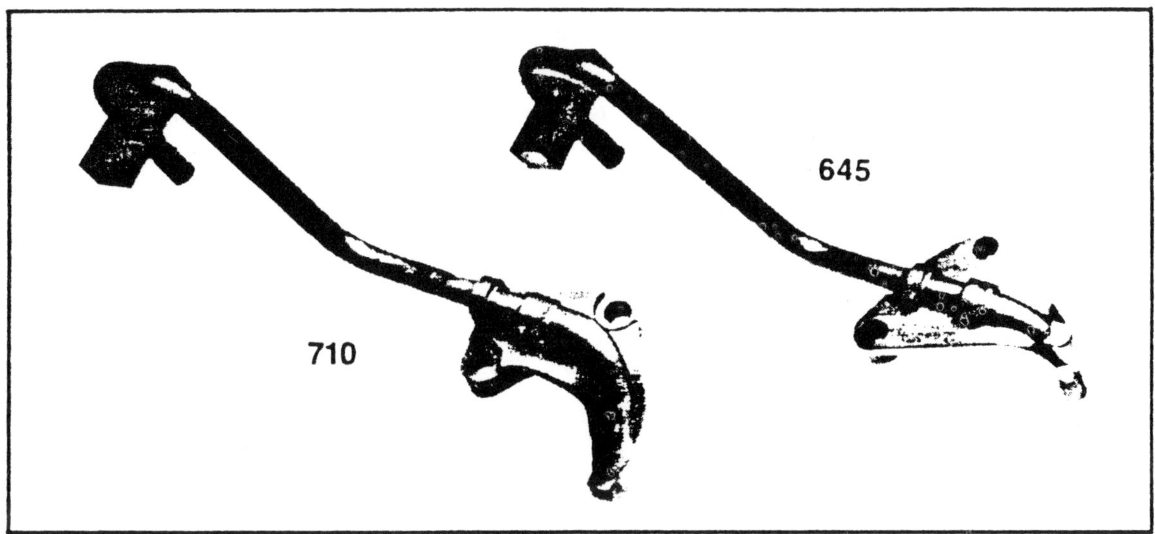

Piston Cooling Tube Assemblies

Maintainability

Maintainability has characterized EMD engines since the initial 567 Series was introduced. The maintainable parts are designed to be as light-weight as possible and are accessible through easily removable covers. Cylinder assemblies and valve train components are readily accessible for inspection without requiring major disassembly.

The 710G power assembly weighs only 450 pounds, 20 pounds more than that of the 645 engine. As with previous EMD engines, power assemblies can be replaced as complete units, thus considerably reducing maintenance time. Torquing procedures are simple and identical to those of 645 engines. Water connections are simple bolted joints; and no intake and exhaust connections are necessary which greatly simplifies maintenance operations compared to that on four-stroke cycle engines of similar size.

New for the 710G engine is a turbocharger overrunning clutch relocated from the interior of the turbocharger to the rear engine gear train. This external clutch can now be serviced independent of the turbocharger thereby reducing both downtime required for inspection or replacement and maintenance costs.

Summary

1. Electro-Motive Division of General Motors has introduced the Model 710G Series Diesel engines. Displacing 710 cubic inches per cylinder as a result of a one-inch increase in stroke from the 645 series, the 16-cylinder 710G engine is rated at 3950 BHP at 900 RPM.
2. Brake specific fuel consumption of the 710G at rated load is ten percent lower than the 1980 model 645F engine and four percent lower than the latest FB model introduced in 1982. The primary performance features are high efficiency turbochargers, a 16:1 compression ratio fire ring piston, a new GM unit injector with 0.5625 inch diameter plunger, and optimized valve and port timing.
3. The 710G engine has conservative mechanical and thermal loadings within previous EMD durability experience on 645 engines.
4. The 710G series continues the simplicity and ease of maintenance of its predecessor, the 645.
5. The 710G Series engines have a high degree of environmental acceptability. Exhaust emissions of rail engines are well below UIC/ORE limits.
6. The increased displacement of 710G Series engine, as well as its conservative initial rating and mechanical and thermal loadings, provide considerable potential for future improvements both in horsepower rating and fuel efficiency.

OPERATOR'S MANUAL

NOTICE

The purpose of this manual is to act as a guide in the operation of the locomotive and its equipment. The information was compiled for a typical locomotive with basic equipment and frequently requested extras. The equipment selected for coverage was chosen as representative and not intended as an indication of availability or use on a particular unit or order.

INTRODUCTION

This manual has been prepared as a guide for railroad personnel engaged in the operation of the 3800 horsepower General Motors SD60 diesel-electric locomotive. "Super Series" locomotives, such as the SD60, are equipped with an advanced electronic wheel control system. Micro-processor equipped locomotives, such as the SD60, are also equipped with a computer display/diagnostic control system.

Locomotive description and operating instructions are divided into four sections as follows:

1. General Description – Describes principal equipment components.
2. Controls – Explains functions of controls used to start and operate the locomotive. Indicating devices to monitor certain locomotive systems also receive coverage.
3. Operation – Outlines procedures for locomotive operation.
4. Indicator Light and Computer Display Messages – Explanation of messages from computer and indicator lights.

To be of most benefit to the reader, these sections should be read in sequence.

Information concerning equipment maintenance, adjustments, and testing is contained in other EMD publications pertaining to this model.

TABLE OF CONTENTS

GENERAL DATA

Model Designation SD60

Locomotive Type . . . . . . (C-C) 0660

Locomotive Horsepower 3800

Diesel Engine
- Model 710G3
- Number of Cylinders 16
- Type . . . . . . Turbocharged
- Full Speed . . . . . . 904 RPM
- Idle Speed . . . . . . 200 RPM

Main Generator Model
- Basic AR11WBA-D18A
- Special AR11A-D14
  - Traction Alternator (AR11A Rectified Output)
    - Maximum Voltage (DC) 1350
    - Continuous Current Rating . . . . . . 7020 Amp
  - Companion Alternator
    - Basic D18
    - Special D14
      - Nominal (AC) Voltage (D18 or D14) 230

AC Auxiliary Generator Voltage (DC) 74
- Maximum Power Output . . . . . . 18 kW

Supplies
- Lube Oil System Capacity
  - Basic Oil Pan . . . . . . 243 Gal.
  - Increased Capacity Pan . . . . . . 395 Gal.
- Cooling System Capacity . . . . . . 276 Gal.
- Sand Capacity (Total)
  - Basic . . . . . . 56 Cu. Ft.
  - Special . . . . . . 72 Cu. Ft.
- Fuel Capacity
  - Basic . . . . . . 3200 Gal.
  - Special . . . . . . 4000 Gal.
  - . . . . . . 4400 Gal.
- Retention Tank . . . . . . 100 Gal.
  - (Reduces fuel capacity by same amount.)

Major Dimensions
- Height (HTC Trucks)
  - Over Cooling Fans . . . . . . 15′ 7 1/8″
  - Over Horn . . . . . . 15′ 5 1/4″
- Width Over Handrail Supports . . . . . . 10′ 3 1/8″
- Length Over Coupler Pulling Faces . . . . . . 71′ 2″
- Approximate Weight on Rails
  - Basic . . . . . . 368,000 lbs.
  - Typically Equipped . . . . . . 390,000 lbs.
- Weight on Drivers 100%

Traction Motors
- Model D87A
- Number 6
- Type . . . . . . DC Series Wound, Axle Hung, Forced Air Ventilated

Maximum Locomotive Speed in MPH
(based on rated RPM of traction motors)

Gear Ratio	40″ Wheel	42″ Wheel
70:17	70	74
69:18	76	80
67:19	82	86
66:20	88	92

Air Compressor
- Model (Basic) . . . . . WLN
  - Type . . . . . 2 Stage
  - Number of Cylinders 3
  - Displacement at 900 RPM . . . . . 254 Cu. Ft./Min.
  - Compressor Cooling . . . . . Engine Coolant
  - Lube Oil Capacity . . . . . 10½ Gal.
- Model (Special) . . . . . WLG
  - Type . . . . . 2 Stage
  - Number of Cylinders 6
  - Capacity (at 900 RPM) . . . . . 400 Cu. Ft./Min.
  - Air Compressor Cooling . . . . . Engine Coolant
  - Lube Oil Capacity . . . . . 18 Gal.
- Model (Special) . . . . . WLO
  - Type . . . . . 2 Stage
  - Number of Cylinders 3
  - Displacement at 900 RPM . . . . . 254 Cu. Ft./Min.
  - Air Compressor Cooling . . . . . Forced Air
  - Lube Oil Capacity . . . . . 10½ Gal.

Air Brakes . . . . . Type 26L

Storage Battery
- Number of Cells 32
- Voltage 64
- Rating (8 Hour) . . . . . 420 Amp Hr.

Minimum Curve Negotiation Capability

Single Unit:
195 Ft. Radius – 29° Curve

Two Coupled Units:
235 Ft. Radius – 24° Curve

Unit Coupled to 50 Ft. Box Car:
379 Ft. Radius – 15° Curve

SECTION 1
GENERAL DESCRIPTION

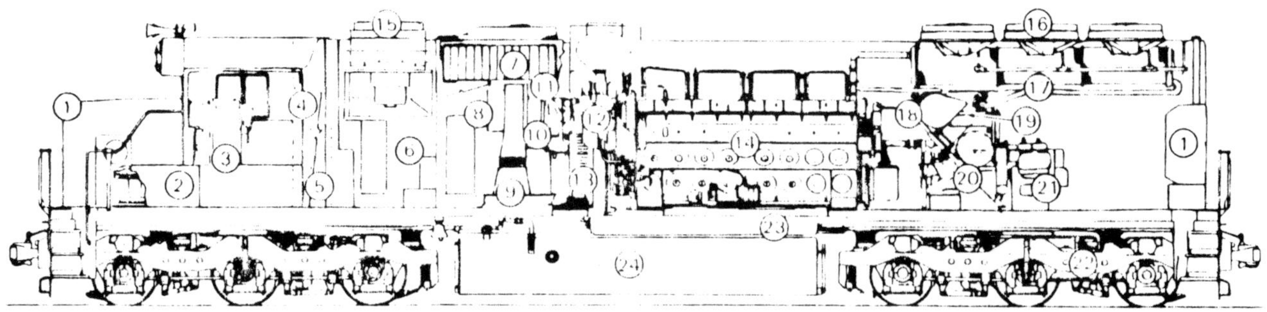

1. Sand Box
2. Battery
3. Control Stand
4. Electrical Cabinet
5. Electric Cabinet Air Filter
6. Generator Transition Contactor
7. Inertial Filter
8. Engine Air Filter
9. Traction Motor Blower
10. Generator Blower
11. Auxiliary Generator
12. Turbocharger
13. Main Generator
14. Diesel Engine
15. Dynamic Brake Blower
16. Radiator Cooling Fans
17. Radiators
18. Engine Water Tank
19. Lube Oil Cooler
20. Lube Oil Filter
21. Air Compressor
22. Truck
23. Main Air Reservoir
24. Fuel Tank

Typical SD60 Locomotive General Arrangement

INTRODUCTION

The General Motors Model SD60 diesel-electric locomotive, Fig. 1-1, is equipped with a turbocharged 16 cylinder diesel engine which drives the main generator. Electrical power from the main generator is distributed to the traction motors through the high voltage control cabinet. Each of the six traction motors is geared directly to a pair of driving wheels. The maximum rated traction motor speed and the gear ratio of the traction motor to the wheel axle determines the maximum operating speed of the locomotive.

The basic locomotive is arranged and equipped so that the short hood or cab end is considered the front or forward part of the unit. However, the locomotive operates equally well in either direction, and on special order controls may be arranged so that the long hood end is forward, or dual controls may be provided.

While each locomotive is an independent power source, several may be combined in multiple operation to increase load capacity. The operating controls on each unit are jumpered or "trainlined" to allow all the locomotives to be simultaneously controlled from the lead unit.

LOCOMOTIVE OPERATION

Storage batteries provide the energy required to start the diesel engine. The engine start switch controls battery power to two starting motor solenoids mounted at the lower rear right hand side of the engine. These electrical solenoids engage the starting motor pinions with the engine ring gear. When both pinions are engaged, battery power is applied to the starting motors to crank the diesel engine.

The diesel engine must be primed with fuel prior to starting. To do this, the operator places the engine start switch in the FUEL PRIME position. This applies battery power to the fuel pump which pressurizes the injector system with fuel. The fuel pump moves the fuel from the fuel tank under the locomotive to the injectors. After the entire system has been supplied fuel, and the injector racks positioned, the cylinder will fire when the engine is cranked. With the engine running, the fuel pump motor is supplied directly by the auxiliary generator.

The diesel engine is the source of locomotive power. When the engine is running, it directly drives three electrical generators and their associated cooling fans, a multi-cylinder air compressor, a traction motor blower, and the water and lube oil pumps. The engine-driven components in the locomotive system must convert the engine power to other forms to perform their individual functions:

1. The main generator rotates at engine speed, generating alternating current power. This power is then converted to direct current power by the internal rectifier banks and directed to the traction motors.

2. The companion alternator is physically coupled to the main generator. It supplies current to excite the main generator field and to power the radiator cooling fans, the inertial filter blower, and various transductors and control devices.
3. The auxiliary generator is driven by the engine gear train at three times engine speed. It provides a 74 volt DC output for excitation current to the companion alternator. The auxiliary generator also supplies the 74 volt power needed for control, cab heating, locomotive lighting, and battery charging circuits.
4. The air compressor, located directly in the engine drive train, supplies the necessary air pressure for brakes and other pneumatic devices such as sanders, windshield wipers, shutter operating cylinders, and horn.
5. The engine gear train drives two centrifugal water pumps which circulate coolant through the engine.
6. The lube oil pumps are also connected in the engine gear train. They supply lubricating oil to critical operating surfaces throughout the engine.

Major components of the diesel-electric power system take power from the diesel engine. The electrical nature of this system is seen in the conversion, application and control of that power.

The main generator supplies electrical energy to the high voltage control cabinet. This cabinet establishes the distribution of power to the traction motors by means of its internal switchgear. The switchgear consists of power contactors, relays, and switches which direct the flow of power as dictated by the control computer. The response of the computer is determined by locomotive operating conditions and the set up of the controls in the cab.

A major part of the locomotive control system involves the interrelated functions of the throttle, governor, and load regulator. The engine governor holds the engine speed at a constant RPM as set by the throttle. It does this by changing the position of the injector racks which control the amount of fuel supplied to each cylinder. Actual operating conditions create varying train loads. When the load changes, the load regulator acts to vary generator excitation. Thus the load regulator balances the governor speed setting from the throttle with the engine power level determined by the load.

As the throttle is advanced to a higher position, the electrical control system causes more current to flow through the field of the main generator. This increased excitation current results in an increase in power to the traction motors. Thus the locomotive power, as well as engine speed, increases progressively in throttle steps.

Most control and protective functions are programmed into the display/diagnostic computer. The computer monitors critical functions in the locomotive power system and provides a display message and audible alarm if a fault occurs.

There are six DC traction motors located on the trucks under the locomotive. Each traction motor is geared directly to the axle on which it is mounted. These motors are supplied power through the high voltage control cabinet at the rear of the cab.

SECTION 2
CONTROLS AND INDICATING DEVICES

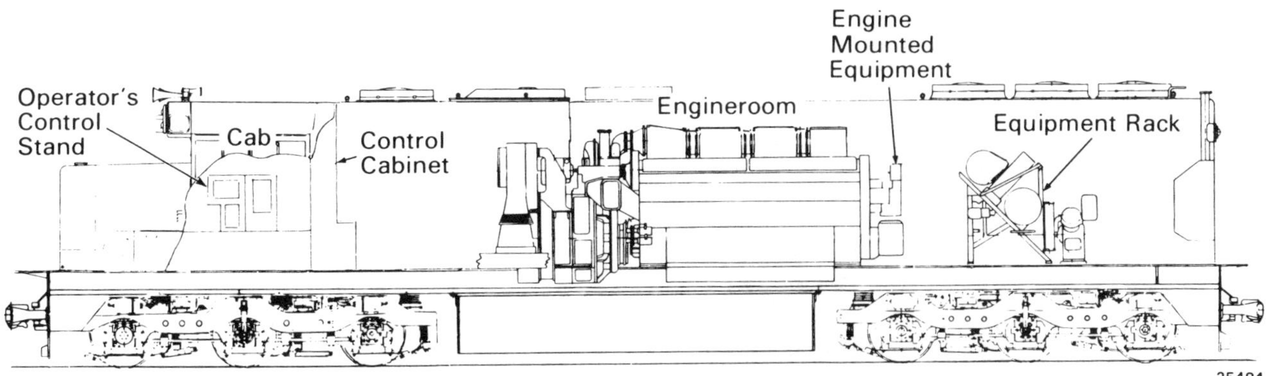

Fig. 2-1 Location of Operating Controls and Indicating Devices

INTRODUCTION

This section provides a brief description of controls and indicating devices used by the operator. Although some equipment receiving coverage is not used during normal operation, it is included to familiarize the operator with its function.

The majority of controls and indicating devices used by the operator are located in the locomotive cab, Fig. 2-1. Engine starting and monitoring equipment is located in the engineroom.

CAB EQUIPMENT

Operating equipment is located in the locomotive cab at two locations: the operator's control stand, and the control cabinet.

OPERATOR'S CONTROL STAND

The operator's control stand, Fig. 2-2, contains switches, gauges, and operating handles used by the operator. The individual components are described, together with their functions, in the following paragraphs.

CONTROLLER

The following operating handles are located on the locomotive controller, Fig. 2-3.

DYNAMIC BRAKE HANDLE

A separate handle is provided for control of dynamic brakes, Fig. 2-4, it is uppermost on the controller panel and is moved from left to right to increase braking effort. The handle grip is somewhat out-of-round with the flattened surfaces vertical to distinguish it from the throttle handle, which has its flattened surfaces horizontal. The brake handle has two detent positions; OFF and SET UP, and an operating range 1 through FULL 8, through which the handle moves freely without notching. Mechanical interlocking prevents the dynamic brake handle from being moved out of the OFF position unless the throttle is in IDLE and the reverser is positioned for either forward or reverse operation.

CAUTION

During transfer from power operation to dynamic braking, the throttle must be held in IDLE for 10 seconds before moving the dynamic brake handle to the SET UP position. This is to eliminate the possibility of a sudden surge of braking effort with possible train run-in or motor flash-over.

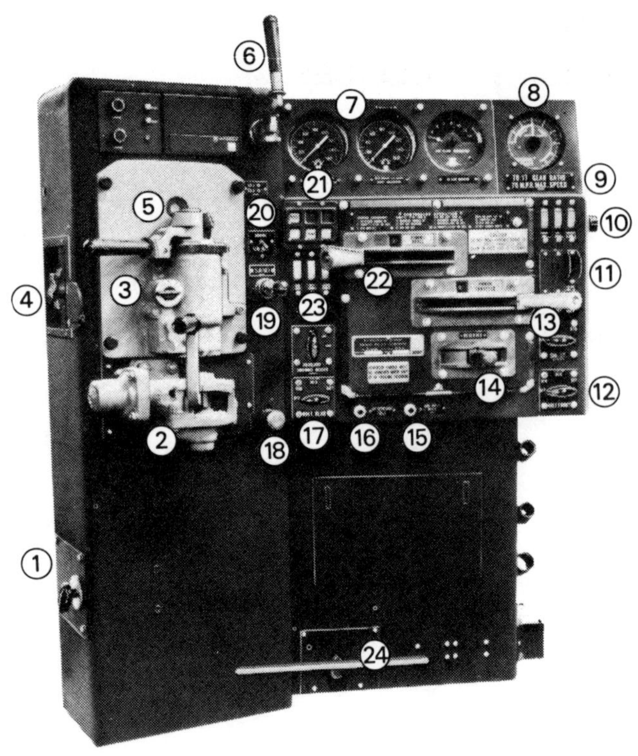

1. Multiple Unit Valve (MU-2A or Dual Ported Cutout Cock)
2. Independent Brake Valve
3. Cut-Off Valve
4. Trainline Air Pressure Adjustment Valve
5. Automatic Brake Valve
6. Air Horn Valve
7. Air Gauges
8. Load Current Indicating Meter
9. Control and Operating Switches
10. Light Dimmer
11. Dynamic Brake Circuit Breaker
12. Headlight Switch-Front
13. Throttle Handle
14. Reverser Handle
15. Signal Light Reset Button
16. Attendant Call Button
17. Headlight Switch-Rear
18. Bell Ringer Valve
19. Manual Sand Lever Switch
20. Lead Truck Sand Switch
21. Indicator Light Panel
22. Dynamic Brake Handle
23. Ground and Light Switches
24. Air Brake Pedal (If Provided)

Fig. 2-2 Typical Operator's Control Stand

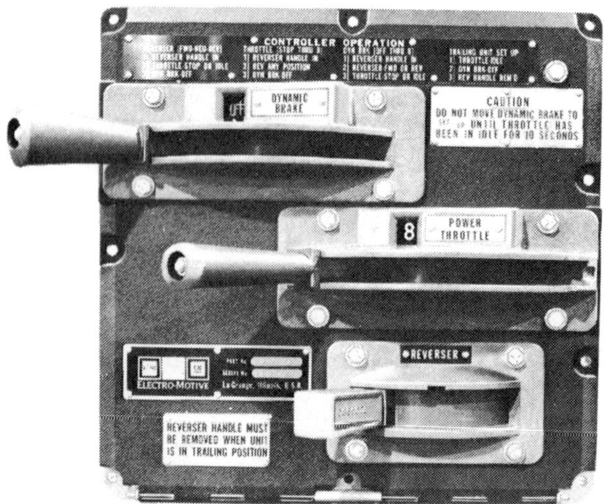

Fig. 2-3 Locomotive Controller

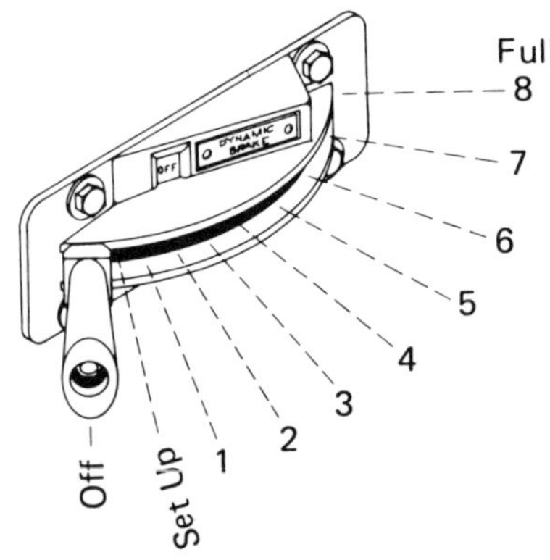

Fig. 2-4 Dynamic Brake Handle

THROTTLE HANDLE

The throttle handle, Fig. 2-5, is located just below the dynamic brake handle. It is moved from right to left to increase locomotive power. The handle grip is somewhat out-of-round, with the flattened surfaces horizontal to distinguish it from the dynamic brake handle. The throttle has nine detent positions; IDLE, and 1 through 8 plus a STOP position, which is obtained by pulling the handle outward and moving it to the right beyond IDLE to stop all engines in a locomotive consist. Mechanical interlocking prevents the throttle handle from being moved out of IDLE into power positions when the dynamic brake handle is advanced to SET UP or beyond, but it can be moved into STOP position to stop all engines in the consist. The throttle can not be moved when the reverser handle is centered and removed from the controller.

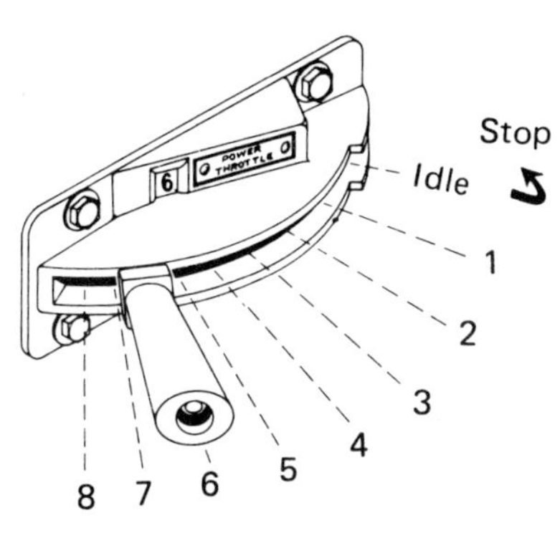

Fig. 2-5 Throttle Handle

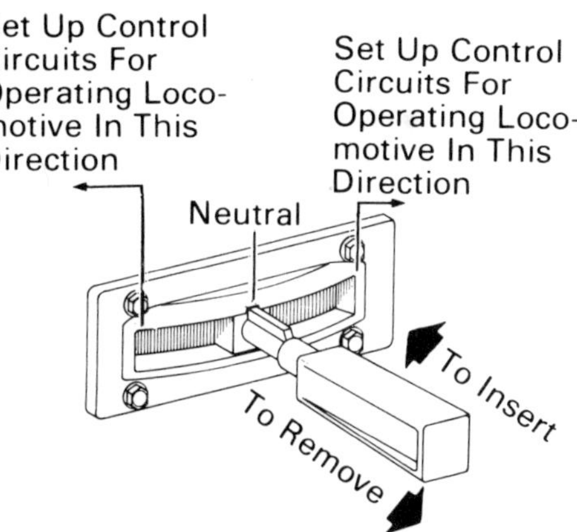

Fig. 2-6 Reverser Handle

REVERSER HANDLE

The reverser handle, Fig. 2-6, is the lowest handle on the controller panel. It has three detent positions; left, centered, and right. When the handle is moved to the right toward the short hood end of the unit, circuits are set up for the locomotive to move in that direction. When the handle is moved to the left toward the long hood end, the locomotive will move in that direction when power is applied. With the reverser handle centered, mechanical interlocking prevents movement of the dynamic brake handle, but the throttle handle can be moved. In such case, power will not be applied to the traction motors.

The reverser handle is centered and removed from the panel to lock the throttle in IDLE position and the dynamic brake handle in OFF position.

MECHANICAL INTERLOCKS ON THE CONTROLLER

The handles on the controller are interlocked so that:

1. With reverser handle in neutral (centered) –
   a. Dynamic brake handle can not be moved out of OFF position.
   b. Throttle can be moved to any position.
   c. Reverser handle can be removed from controller if throttle is in IDLE position.

2. Reverser handle in forward or reverse –
   a. Throttle can be moved to any position if dynamic brake handle is in OFF position.
   b. Dynamic brake handle can be moved to any position if throttle is in IDLE position.

3. Reverser handle removed from controller –
   a. Throttle locked in IDLE position.
   b. Dynamic Brake handle locked in OFF position.

4. Throttle in IDLE position –
   a. Dynamic brake handle can be moved to any position if reverser is in forward or reverse position.
   b. Reverser handle can be placed in neutral, forward, or reverse position if dynamic brake handle is in OFF position.

5. Throttle above IDLE position –
   a. Dynamic brake handle can not be moved.
   b. Reverser handle can not be moved.

6. Dynamic brake handle in OFF position –
   a. Throttle can be moved to any position.
   b. Reverser handle can be moved to any position if throttle is in IDLE position.

7. Dynamic brake handle moved out of OFF position –
   a. Throttle can not be moved out of IDLE position into power positions, but can be moved into STOP position.
   b. Reverser handle can not be moved out of forward or reverse into OFF position.

26L AIR BRAKE EQUIPEMENT

Basic locomotives are equipped with type 26L air brake equipment. This equipment is located to the left of the controller and as shown in Fig. 2-7 includes an automatic brake, independent brake, cut-off pilot valve, a trainline air pressure adjustment valve, and either a dual ported cutout cock or an MU-2A valve.

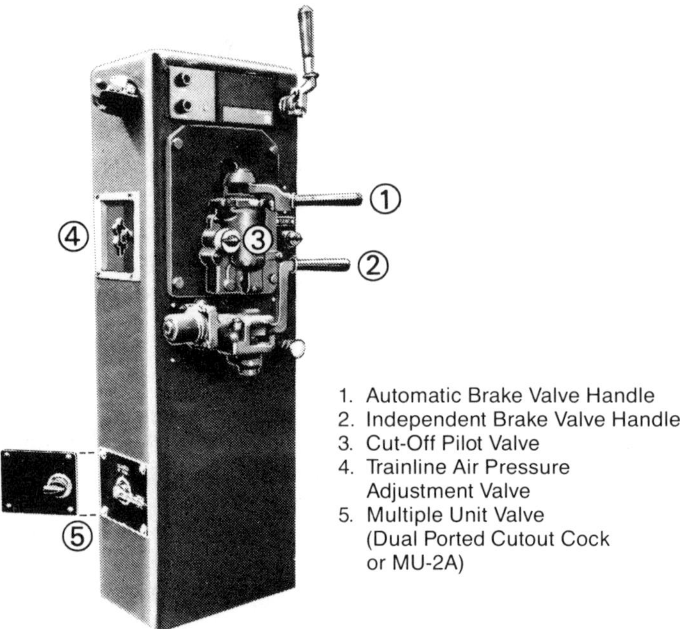

Fig. 2-7 Typical Air Brake Equipement

A dead engine feature is also part of the 26L air brake equipment. The dead engine cutout cock and pressure regulator, Fig. 2-8, are accessible from outside the locomotive through side doors provided. The pressure regulator is set by maintenance personnel and is not to be set by the operator.

AUTOMATIC BRAKE VALVE HANDLE

The automatic brake valve handle, Fig. 2-9, controls the application and release of both the locomotive and train brakes. The brake valve is of the "pressure maintaining type" which will hold brake pipe reductions constant against nominal brake pipe leakage. A brief description of the operating positions follows:

Release Position

This position is for charging the equipment and releasing the locomotive and train brakes. It is located with the handle at the extreme left of the quadrant.

Minimum Reduction Position

This position is located with the handle against the first raised portion on the quadrant to the right of release position. With the handle moved to this position, minimum braking effort is obtained.

Service Zone

This position consists of a sector of handle movement to the right of release position. In moving the handle from left to right through the service zone, the degree of braking effort is increased until, with the handle at the extreme right of this sector, the handle is in full service position and full service braking effort is obtained.

Suppression Position

This position is located with the handle against the second raised portion of the quadrant to the right of release position. In addition to providing full service braking effort, as with the handle in full service position, suppression of overspeed control and safety control application, if equipped, is obtained.

Handle Off Position

This position is located by the first quadrant notch to the right of suppression position. If so equipped, the handle is removable in this position. This is the position in which the handle should be placed on trailing units of a multiple-unit locomotive or on locomotives being towed "dead" in a train.

Emergency Position
This position is located to the extreme right of the brake valve quadrant. It is the position that must be used for making brake valve emergency brake applications and for resetting after any emergency application.

INDEPENDENT BRAKE VALVE HANDLE
The independent brake valve handle, Fig. 2-10, is located directly below the automatic brake handle. This handle provides independent control of the locomotive braking effort irrespective of train braking effort. The brake valve is self-lapping and will hold the brakes applied. A brief description of the operating positions follows.

Release Position
This position is located with the handle at the extreme left of the quadrant. This position releases the locomotive brakes, provided the automatic brake handle is also in release position.

Full Application Position
This position is located with the handle at the extreme right of the quadrant. In moving the handle from left to right through the service zone the degree of locomotive braking effort is increased until full application braking effort is obtained.

Depression of the independent brake handle whenever the handle is in release position will cause the release of any automatic brake application existing on the locomotive. Depression of the independent brake handle when in the service zone will release the automatic application of the locomotive brakes to the value corresponding to the position of the independent brake handle.

NOTE

This locomotive model can be equipped with either an MU-2A valve, or a dual ported cutout cock for multiple unit control. Information about both of these devices is provided.

DUAL PORTED CUTOUT COCK
The dual ported cutout cock (if so equipped), Fig. 2-11, is located at the lower left side of the control stand. Its purpose is to set up the locomotive brake system for lead, trail, or dead operation. The handle is placed in the CLOSED IN TRAIL position when the unit is trailing in a consist, and is placed in the OPEN IN LEAD OR DEAD position when leading or dead.

MU-2A VALVE
The MU-2A valve (if so equipped), Fig. 2-11, is located on the left hand side of the air brake stand. Its purpose is to pilot the F1 selector valve which is a device that enables the air brake equipment of one locomotive unit to be controlled by that of another unit.

The MU-2A valve has two positions which are:

1. LEAD OR DEAD   2. TRAIL 24 OR 26

The valve is positioned by pushing in and turning to the desired setting.

CUT-OFF PILOT VALVE
The cut-off pilot valve, Fig. 2-7, is located on the automatic brake valve housing directly beneath the automatic brake handle. The valve has the following two positions:

1. OUT   2. IN

To operate locomotive as the controlling unit, the cut-off valve handle must be pushed in and rotated to the IN position. The OUT position is used when hauling the locomotive "dead" or as a trailing unit in a consist.

On special order the cut-off pilot valve may have the following three positions:

1. OUT
2. FRT (freight)
3. PASS (passenger)

In this case the valve is pushed in and placed in the position desired, depending on make-up of train.

TRAINLINE AIR PRESSURE ADJUSTMENT VALVE
The trainline air pressure adjustment valve, Fig. 2-7, is located to the left of the automatic brake valve. With the automatic brake valve handle in release position, it is used to obtain the brake pipe pressure desired. The automatic brake valve will maintain the selected pressure against overcharge or leakage.

26L AIR BRAKE EQUIPMENT OPERATING POSITIONS

In the absence of specific instructions, usually issued by each railroad to cover its own recommended practices, refer to Fig. 2-12 for brake equipment operating positions most often encountered while the locomotive is in service.

Type Of Service	Automatic Brake Valve	Independent Brake Valve	Cutoff Valve	Dead Engine Cutout Cock	26F Control Valve	Multiple Unit Valve		Overspeed Cutout Cock	Alertor Cutout Cock	Deadman Cutout Cock
						MU-2A Valve	Dual Ported Cutout Cock			
SINGLE LOCOMOTIVE EQUIPMENT										
Lead	Release	Release	In*	Closed	Graduated Direct	Lead	Open	Open	Open	Open
Shipping Dead In Train	Handle Off Position	Release	Out	Open	Direct	Dead	Open	Closed	Closed	Closed
MULTIPLE LOCOMOTIVE EQUIPMENT AND EXTRAS										
Lead	Release	Release	In*	Closed	Graduated Direct	Lead	Open	Open	Open	Open
Trail	Handle Off Position	Release	Out	Closed	Graduated Direct	Trail 24 or 26	Closed	Open	Open	Open
Shipping Dead In Train	Handle Off Position	Release	Out	Open	Direct Release	Dead	Open	Closed	Closed	Closed

* On units equipped with a three position cut-off valve, position valve to either FRT or PASS, depending on make-up of train.

Fig. 2-12 26L Air Brake Equipment Positions

MISCELLANEOUS CONTROLS AND SWITCHES

The following paragraphs describe miscellaneous controls, switches, and indicators typically provided on the operator's control stand, Fig. 2-2.

AIR HORN VALVE

When the air horn lever is pulled, compressed air is supplied to the locomotive air horn.

SANDING SWITCHES

Manual sand is cut out when the locomotive is in Super Series operation and moving above 5 mph. However, if a Super Series locomotive is in consist with older units, movement of the sand lever switch will supply a trainlined signal to the older units and sand will be applied.

Sanding Lead Truck Toggle Switch

The signal from this switch is not trainlined. The switch provides sand to only the lead truck. This method of sanding dresses the rail and is adequate for most conditions. The SAND light will come on when this switch is activated.

Sand Lever Switch

When operated, this lever supplies a signal to the sanding input of the control computer. The signal also causes the SAND light on the operator's control stand to turn on. The computer determines which direction the locomotive is moving and directs the trainlined signal to the appropriate (forward or reverse) sanding magnet valves. The basic switch is non-latching and may be operated in any direction for correct sanding. A directional sanding switch may be provided as an optional extra, and the switch may be latching if requested by the railroad.

Electrically controlled sanding is the basic system used, but since the locomotive may be operated in multiple with older units that are equipped only for pneumatic control of sanding, trainlined pneumatic control of sanding may be provided as an optional extra in addition to electrical control. In such cases, trainlined actuating pipes must be connected between units.

BELL RINGER VALVE

This mushroom type valve actuator operates the locomotive signal bell.

INDICATOR LIGHT PANEL

The indicator light panel, Fig. 2-13, contains lights to indicate operation of various systems within the locomotive. The panel has provisions for six press-to-test lights covered by either white or colored lens caps identified in black letters. The lights are discussed in the following paragraphs.

NOTE

The following indicator lights have a push-to-test feature which allows testing of the lamp circuit alone, isolated from its operation in the power control system. When the lens cap is depressed the supply voltage is impressed across the lamp circuit. After a one second delay the light should go on.

WHEEL SLIP Light

The wheel slip light comes on to indicate one of the following four conditions:

1. Slipping wheels while starting a train. When starting a train, the starting wheel slip system functions to correct wheel slips. Intermittent flashing of the wheel slip light indicates moderate to severe wheel slip. The throttle (locomotive power) should not be reduced unless severe lurching threatens to break the train.

NOTE

Minor slips or wheel creep will not actuate the wheel slip light, but automatic sanding may take place along with regulation of power to the wheels. Do not misinterpret this power control as loss of power due to a fault.

2. Wheel Overspeed. Overspeed conditions which may result from simultaneous wheel slip or excessively high track speed will activate the wheel slip light. In each case locomotive power will be regulated automatically to correct the condition. This condition is indicated when the wheel slip light *cycles on and off.*

CAUTION

Irregular flashing of the wheel slip light when the locomotive is in power and above 1.5 mph could be an indication of a Super Series failure. Operation may continue, but the condition must be reported to authorized maintenance personnel.

3. Locked powered wheel. A locked wheel condition will cause the wheel slip light to be on if the locomotive is under power. This condition is indicated when the wheel slip light is *on continuously.*

WARNING

Never operate the locomotive with a continuous wheel slip light (locked wheel condition). If circuit difficulty is suspected, stop the locomotive and make a careful inspection to ascertain that there are no locked sliding wheels before proceeding.

4. Wheel slips during dynamic braking. When equipped for dynamic braking, the wheel slip light will come on to indicate when a pair of wheels is detected tending to rotate at a slower speed.

PCS OPEN Light

The PCS OPEN light comes on to indicate a safety control or emergency air brake application. The pneumatic control switch PCS functions to automatically cut power to the traction motors in the event of a safety control or emergency air brake application.

Locomotive power is restored by pick up of the pneumatic control relay PCR. This occurs automatically, provided that:

1. Control of the air brake is recovered (PCS resets).
2. The throttle is returned to IDLE position.
3. The P.C. RESET pushbutton switch (if so equipped) on the control stand is operated.

NOTE

If the locomotive is not equipped with a P.C. RESET switch on the control stand, then only the first two conditions are required to restore tractive power.

BRAKE WARN Light

This light indicates excessive dynamic braking current. In the event that the brake warning light comes on, reduce dynamic brake handle position immediately to decrease braking effort and prevent possible equipment damage. If the brake warning light does not go out or if the indication repeats, place the dynamic brake cutout switch on the engine control panel of the affected unit in the CUTOUT position. The unit will then operate normally under power, but not in dynamic braking. Total dynamic braking effort of units coupled in consist will be reduced.

OSC HDLT Light

Comes on to indicate that the red oscillating headlight is on, either because of emergency or penalty control brake application or because the signal light control switch on the control stand is positioned at RED.

SAND Light

This light indicates that one of the manual sanding switches is activated.

NOTE

Automatic sanding, initiated by locomotive control circuits, or emergency sanding will not cause the sand light to come on.

S.C. NOT OPER. Light

Comes on to indicate that the automatic train speed control system is not operating. This occurs with either the mode selector switch on the cab indicator in OFF position or the speed control switch on the controller mounted speed control switch box in the OFF position.

LIGHT SWITCHES

Switches for the ground/step light and gauge lights are located to the left of the controller. The lights are on when the switches are in the up position.

7TH THROTTLE KNOCKDOWN SWITCH

This switch, when so equipped, causes 8th throttle operation to be reduced to 7th throttle engine speed and power when placed in the down position.

HEADLIGHT SWITCHES

Two four-position rotary snap switches are provided for independent control of the front and rear headlights. Each switch has OFF, DIM, MED, and BRT. positions.

CONTROL AND OPERATING SWITCHES

A group of three operating switches, Fig. 2-14, is located at the upper right corner of the control stand. They snap into the on position when moved upward. The switches must be set in the on position when the unit leads a consist, and must be set in the off position in trailing consist units.

Engine Run Switch

This switch must be on to obtain throttle control of engine speed. If the engine run switch is off, the engine will run at idle speed regardless of throttle handle position.

Gen. Field Switch

The generator field switch must be on to complete the excitation circuits to the main generator. If the switch is in the off position, the engine will respond to throttle, but the generator will not develop power.

Control & Fuel P. Switch

The control and fuel pump switch provides power to various low voltage control circuits. The switch must be on to start the engine and operate the fuel pump.

SIGNAL LIGHT SWITCH

This switch is used to turn on the locomotive signal light.

DYNAMIC BRAKE CONTROL CIRCUIT BREAKER

On locomotives equipped for dynamic braking, this circuit breaker is provided to protect against a faulty operating or test setup. The circuit breaker should be in the on (up) position for normal operation. A tripped circuit breaker generally indicates that during dynamic brake testing more than one dynamic brake handle in a locomotive consist was out of OFF position.

ATTENDANT CALL PUSHBUTTON

When this button is pressed in any unit coupled in consist, the alarm bell will ring in all units.

P.C. RESET Pushbutton Switch

This pushbutton switch (if so equipped) must be operated to restore locomotive power in the event of an emergency or safety control air brake application.

SPOTTER Pushbutton Switches

The spotter circuit (if so equipped) is used to position the locomotive a slight distance by applying battery power to two traction motors in series. Both switches must be operated at the same time.

NOTE

The locomotive is equipped with an automatic ground relay reset function. This computer function will automatically reset the ground relay a certain number of times in a time period before it locks out. Refer to ground relay messages in Section 4. A ground relay lockout reset switch is located onthe No. 2 circuit breaker panel of the electrical cabinet. Refer to railroad regulations before resetting the lockout function.

AIR GAUGES

Air gauges to indicate main reservoir air pressure as well as various pressures concerned with the air brakes are located along the top of the control stand.

LOAD CURRENT INDICATING METER

Locomotive pulling force is indicated by the load current indicating meter. The meter is graduated to read amperes of electrical current, with 1650 being the maximum on the scale. A red area on the meter face indicates when the current levels are too high for continuous operation. The meter is connected to indicate average traction motor current.

CAUTION

Observe short time operation plate instructions pertaining to low speed full throttle operation. This plate is located below the load current indicating meter.

Maximum continuous current rating and the short time operating limits were developed for throttle 8 operation. These values must be decreased at lower throttle positions because engine speed, and, consequently traction motor cooling air are reduced.

On locomotives equipped for dynamic braking, a zero-center type meter is applied, Fig. 2-15. The meter needle swings to the right of zero to indicate load current during power operation, and it swings to the left of zero to indicate dynamic braking current, with 900 amperes being the maximum reading on the braking portion of the meter.

Since the dynamic brake regulator controls maximum braking current, the meter should seldom, if ever, indicate more than 760 amperes, which is the basic rating of the dynamic braking resistor grids.

POWER REDUCTION CONTROLS (If Provided)

These controls, Fig. 2-16, if provided, are located on the operator's control stand.

NOTE

A 5 mph limit on power reduction is built into the control system. On special order, the 5 mph limit is deleted.

POWER REDUCTION INDICATOR LIGHT

The power reduction indicator light is on only when the power reduction toggle switch is in the LOCAL or the TRAINLINE position, which means that the setting of the power reduction rheostat on the same unit is affecting traction power. The indicator light is off when the toggle switch is OFF. See previous NOTE.

POWER REDUCTION TOGGLE SWITCH

When the power reduction toggle switch is in the LOCAL position, the power reduction rheostat on that unit controls power reduction only on that unit. However, the LOCAL position of the toggle switch does not prevent power reduction on that unit from being controlled by another unit's power reduction circuit.

When the power reduction toggle switch is in the TRAINLINE position, the power reduction rheostat on that unit controls power reduction on that unit and on the other units in the consist.

NOTE

If the power reduction toggle switch on any unit of a consist is in the TRAINLINE position, the power reduction toggle switches on all the other consist units should be in the OFF position. See previous NOTE.

When the power reduction toggle switch is in the OFF position, the power reduction rheostat on that unit has no effect. If the unit is part of a consist, the OFF position of the switch prevents that unit's power reduction rheostat from interfering with the controlling unit.

POWER REDUCTION RHEOSTAT

When activated by the LOCAL or the TRAINLINE position of the power reduction toggle switch, the power reduction rheostat overrides the power level called for by the throttle. For example, if the rheostat is set to reduce power by 1/3, it will do so in each throttle position. When set in the MAX position, fully clockwise, the rheostat permits normal power in each throttle position.

CONTROL CABINET

The control cabinet, Fig. 2-17, contains an engine control panel, a fuse and switch panel, three circuit breaker panels, and the display/diagnostic panel, Fig. 2-18, in the center door. Each panel contains controls and/or indicating devices used by the operator

WARNING

Never open any control cabinet doors other than to gain access to the circuit breaker and fuse and switch panels. High voltage and current are present throughout the control cabinet.

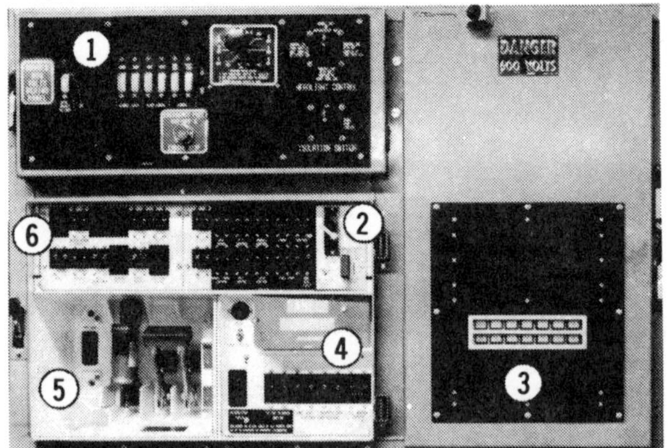

1. Engine Control Panel
2. No. 2 Circuit Breaker Panel
3. Computer Display / Diagnostic Panel
4. No. 3 Circuit Breaker Panel
5. Fuse and Switch Panel
6. No. 1 Circuit Breaker Panel

Fig. 2-17 Typical SD60 Control Cabinet Panels

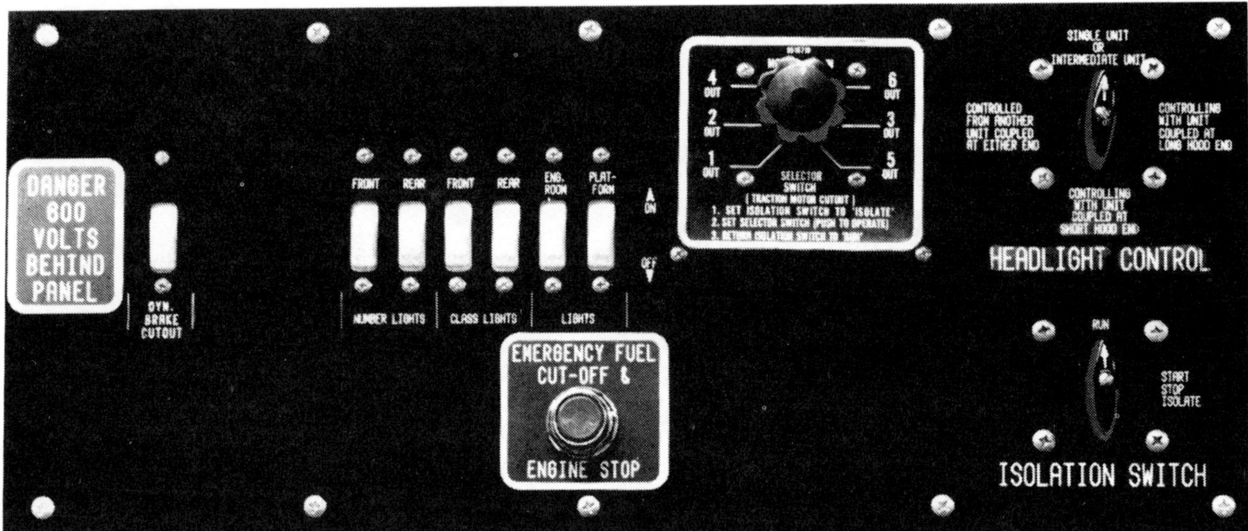

Fig. 2-19 Engine Control Panel, With Typical Extras

ENGINE CONTROL PANEL

The engine control panel, Fig. 2-19, contains various switches used in the operation of the locomotive. A brief description of each switch is provided.

EMERGENCY FUEL CUT-OFF & ENGINE STOP SWITCH

The diesel engine will stop whenever the engine stop pushbutton is pressed. The reaction to the pushbutton is immediate and it need not be held in until the engine stops.

TRACTION MOTOR CUTOUT SWITCH

The traction motor cutout switch, if provided, operates to electrically isolate a defective traction motor. This permits operation with the remaining good motors. The power control system automatically limits power to prevent overloading the operative motors. The isolated motor will continue to rotate as the train moves.

Observe instructions printed on the panel when necessary to cut out a traction motor.

WARNING

Make certain that *all wheels rotate freely* before operating with a motor cut out.

The dynamic brake system is disconnected when operating with a motor cut out.

HEADLIGHT CONTROL SWITCH

The twin sealed-beam front and rear headlights are controlled by the front and rear headlight switches on the locomotive control stand. Before these switches will function, the headlight circuit breaker must be placed on.

On the locomotives equipped for multiple unit operation, a remote headlight control switch is mounted on the engine control panel. This remote headlight control switch provides for operation of the rear unit headlight from the lead unit. The switch positions are set on each unit as follows:

On Lead Unit

If only a single locomotive unit is being used, place the switch in SINGLE UNIT position.

In multiple unit service, if trailing units are coupled to the No. 2 or long hood end of the lead unit, place the switch in the CONTROLLING WITH UNIT COUPLED AT LONG HOOD END position.

In multiple unit service, if trailing units are coupled to the No. 1 or short hood end of the lead unit, place switch in CONTROLLING WITH UNIT COUPLED AT SHORT HOOD END position.

On Intermediate Units

On units operating in between other units in a multiple unit consist, place the switch in the INTERMEDIATE UNIT position.

On Trailing Units

The last unit in a multiple unit consist should have the headlight control switch placed on CONTROLLED FROM ANOTHER UNIT COUPLED AT EITHER END.

ISOLATION SWITCH

The isolation switch has two positions, one labeled START/STOP/ISOLATE, the other labeled RUN. The functions of these two positions are as follows:

START/STOP/ISOLATE Position

The isolation switch is placed in this position whenever the diesel engine is to be started. The start switch is effective only when the isolation switch is in this position.

This position is also used to isolate the unit, and when isolated the unit will not develop power or respond to the controls. In this event the engine will run at idle speed regardless of throttle position. This position will also silence the alarm bell in the event of a no power or low lube oil alarm. It will not, however, stop the alarm in the event of a hot engine.

If the locomotive is equipped with the remote traction motor cutout switch feature, the isolation switch must be placed in the ISOLATE position before the cutout switch can be operated.

RUN Position

After the engine has been started, the unit can be placed "on the line" by moving the isolation switch to the RUN position. The unit will then respond to control and will develop power in normal operation.

THROTTLE LIMIT SWITCH

When equipped, this switch allows locomotive power to be reduced when operating in throttle positions 7 or 8, while trailing units remain at normal power. Reduced lead unit power helps dress the rails for trailing units. In throttle positions 7 or 8 power will be reduced to the equivalent throttle position 5 or 6 respectively, when this switch is in the THROTTLE LIMIT position. If operating in throttle position 6 or below, power will not be reduced.

LOCKED WHEEL CUTOUT SWITCH (If Provided)

For the locked wheel detection system to be operational, this switch must be in the LOCKED WHEEL (up) position. The locked wheel detection system is then effective whether the unit is under power, is shut down, is isolated, or has motors cut out. When the switch is in CUTOUT position, locked wheel detection is nullified; however, the wheel slip control system will provide protection against a locked wheel on any units under power without motors cut out.

Should a temporary operating condition such as unequal release of air brakes bring about a locked wheel indication, automatic reset will occur when the wheel again turns freely. When a locked wheel indication is received, follow the procedure outlined under the #N AXLE LOCKED display message in Section 4.

DYNAMIC BRAKE CUTOUT SWITCH (If Provided)

On units so equipped, when this switch is placed in the CUTOUT position, the individual unit will not operate in dynamic braking. It will however, continue to operate normally under power. The switch can be used to limit the number of units coupled together that will operate in dynamic braking, or it may be used to cut out a unit that is defective in dynamic braking, yet allow it to operate under power.

MISCELLANEOUS LIGHT SWITCHES

Switches are included in circuits for various lights on the locomotive. The switches are closed as desired to operate the class lights, number lights, engineroom lights, and platform lights.

CIRCUIT BREAKER PANELS

The three circuit breaker panels contain circuit breakers and controls used to protect engine, control systems, lights and miscellaneous devices that are used as conditions require. These circuit breakers can be operated as switches, but will trip open when an overload occurs.

NO. 1 CIRCUIT BREAKER PANEL

This panel contains circuit breakers that protect customer requested extra equipment. The No. 1 circuit breaker panel, Fig. 2-20, has provisions for twelve circuit breakers. The following paragraphs contain a brief description of typical circuit systems protected by breakers on this panel.

Water Drain

This breaker protects the automatic cooling system drain circuit, if circuit is provided. It must be closed to enable arming the circuit (see FUEL PRIME / ENGINE START switch) and must then remain closed continuously to enable the circuit to provide protection against freezing.

Radio Control

When equipped for remote radio control this breaker protects radio control circuits.

Air Cond. Blower

When equipped with air conditioning this breaker protects the blower fan motor circuits. A separate breaker for the air conditioner compressor is located on No. 3 circuit breaker panel.

Aux. Cab Htr.

These breakers protect the left and right auxiliary cab heaters. Heat control is provided by switches located on the control stand or at the heater.

Radio

Protects circuits that supply the radio, when equipped.

Utilities

When equipped, this breaker protects the toilet immersion heater, or similar devices.

Auto. Drain Timer

Protects circuits that control automatic operation of drain valves in the compressed air system.

Warning Devices

This breaker protects signal light circuits, when equipped. This breaker may also be used to protect similar devices.

Safety Devices

Train overspeed brings about a penalty application of the brakes and operation of the pneumatic control switch to drop locomotive power. This breaker protects the overspeed magnet valve circuit. This breaker may also be used to protect similar devices.

NO. 2 CIRCUIT BREAKER PANEL

The No. 2 circuit breaker panel, Fig. 2-21, contains circuit breakers and switches that protect basic locomotive equipment and control systems. The panel is divided into three sections. The shaded middle section indicates breakers required on for locomotive operation. Breakers in the unshaded section are used as conditions require.

Gen. Transition

This breaker protects the main coil of the main generator series contactor, and a portion of its control circuit. Closing it enables forward (parallel-to-series) main generator transition, for high speed operation.

A.C. Control

This breaker protects that portion of the control system equipment that is supplied 230 VAC power from the companion alternator (D14/D18). This includes part of the ground relay circuit and feedback devices for armature, grid, and main generator field current. If the breaker trips during locomotive operation, the main generator will not develop power and the NO LOADING-NO D18 OUTPUT display message will come on indicating no companion alternator output.

Module Control
This breaker protects the local control circuit that supplies power to the circuit modules and miscellaneous control system devices.

Turbo
This breaker must be in the on position to start the engine and operate the turbocharger auxiliary lube oil pump. It must remain in the on position to provide auxiliary lubrication to the turbocharger at engine start and after the engine is shut down.

Control
This breaker sets up the fuel pump and control circuits for engine starting. Once the engine is running, power is supplied through this breaker from the auxiliary generator to maintain operating control.

Brake Trans. Control
This double pole breaker is located in the feed to the operating motor of the multi-pole, motor operated, ganged switches that control the motor field and armature connections for either dynamic braking or power operation. Since control power is required to move the transfer switchgear from any position to any other position, the breaker must be closed for power transfer to take place. An open breaker does not prevent switchgear that is already in position from conducting traction motor current, but interlocking prevents an operating setup in conflict with transfer switch position.

Rev. Control
This breaker is located in the feed to the operating motor of the multi-pole, motor operated, ganged switches that control the direction of current flow through the traction motor fields and thus control the direction of locomotive travel. Since control power is required to move the RV transfer switchgear from any position to any other position, this breaker must be closed for power transfer to take place. An open breaker does not prevent switchgear that is already in position from conducting traction motor current, but interlocking prevents an operating setup in conflict with transfer switch position.

Local Control
This circuit breaker establishes "local" power from the auxiliary generator to operate heavy duty switchgear and various control devices.

Fuel Pump
This breaker protects the fuel pump motor circuit. A fuel filter bypass valve may be provided to prevent overloading the fuel pump motor if the fuel filter becomes clogged.

Aux. Gen. Field
The field excitation circuit of the auxiliary generator is protected by this breaker. In the event that this breaker trips, it stops auxiliary generator output to the low voltage sytem and also stops fuel pump operation. An alternator failure (no power no battery charge) alarm occurs. The engine will stop from lack of fuel.

Lights
This breaker must be on to supply power to switches that control miscellaneous locomotive lights.

Hdlts.
This breaker must be on to provide current to the front headlight circuit and through the trainline to the light at the rear of the consist.

Ground Relay Cutout Switch
The purpose of the ground relay cutout switch is to eliminate the ground protective relay from the locomotive circuits during certain shop maintenance inspections. It must always be kept closed in normal operation. When this switch is open, it prevents excitation of the main generator and throttle response of the diesel engine in addition to cutting out the ground protective relay.

Ground Relay Lockout Reset Switch
A ground relay pickup will be reset automatically by the computer. The computer will also lower the main generator operating voltage a fixed amount for each repeated ground relay pickup in a time period. When this occurs a VOLTAGE LIMITING DUE TO GROUND RELAY message will be displayed.

If the number of ground relay pickups per unit of time does not become excessive, then main generator voltage will gradually be increased until normal operating voltage and full power is restored.

If the number of ground relay pickups per unit of time becomes excessive, then the computer will lock out the ground relay which disables the locomotive. Refer to ground relay messages in Section 4. The ground relay lockout reset switch may be used to restore operation. Refer to railroad regulations before resetting the lockout function.

Engine Purge Bypass Switch

This switch is used to bypass the engine purge function in the event that there is a failure in that system. Refer to instruction nameplate on electrical cabinet door.

NO. 3 CIRCUIT BREAKER PANEL

The No. 3 circuit breaker panel, Fig. 2-22, has provisions for five circuit breakers. The panel also contains a sealed section. This section contains a test panel intended for use by maintenance personnel during maintenance and testing procedures. A 74 volt receptacle and fuse test switch are also part of this panel.

The circuit breaker portion of the panel is divided into two sections. Breakers in the shaded section are required on for locomotive operation. Breakers in the unshaded sectio are to be used as conditions require.

Generator Field

The main generator receives excitation current through a controlled rectifier from the companion alternator. This breaker is provided to protect the controlled rectifier and both generators as well as associated circuitry.

NOTE

Unlike other breakers on the panel that trip to the full position, the generator field circuit breaker will trip to the center position. After a period for cooling, the breaker must be placed in the full off position before resetting to the on position.

Filter Blower Motor

This breaker protects the inertial filter blower motor circuit. The blower is used to evacuate dirt loaded air from the central air compartment inertial filters.

The FILTER BLOWER MOTOR CIRCUIT BREAKER DOWN message will be displayed if this breaker trips open or is inavertently left in the off position. If tripped open, operation may continue to the nearest maintenance point.

Electric Cab Heaters

Eng. Side

Protects circuits to the cab heater at the operator's station.

Helper's Side

Protects circuits to the cab heater at the helper's side of the cab.

Air Cond. Comp.

When equipped with air conditioning, this breaker protects the air conditioner compressor motor circuits. A separate breaker for the air conditioner blower fan motor is located on the No. 1 circuit breaker panel.

A.C. Water Cooler

When equipped with a water cooler powered by AC voltage from the companion alternator (D14/D18), this 15 A circuit breaker protects the water cooler motor.

Fuse Test Switch

Refer to Fuse Test Equipment paragraph under the Fuse and Switch Panel section.

74 Volt Receptacle

This receptacle makes 74 volts DC available for maintenance or testing purpose. Power is supplied to the receptacle when the main battery switch and the LIGHTS circuit breaker are closed.

FUSE AND SWITCH PANEL

The fuse and switch panel, Fig. 2-23, contains the equipment described in the following paragraphs.

NOTE

There is no companion alternator field fuse. If a fault occurs in this circuit, auxiliary generator voltage will come down, and the machine will not be harmed. A NO LOADING – NO D18 OUTPUT and a NO LOADING – NO AUX GEN OUTPUT display message will be seen, accompanied by an audible alarm.

AUXILIARY GENERATOR CIRCUIT BREAKER

This breaker connects the auxiliary generator to the low voltage system. It protects against excessive current demands.

NOTE

Unlike other breakers that trip to the full off position, this breaker will trip to the center position. After a period for cooling, the breaker must be placed in the full off position before resetting to the on position.

STARTING FUSE

The starting fuse is in use only during the period that the diesel engine is actually being started. At this time, battery current flows through the fuse and starting contactor to the starting motors.

Although this fuse should be in good condition and always left in place, it has no effect on locomotive operation other than for engine starting. A defective fuse can be detected when attempting to start the engine, since at that time (even though the starting contactors close) the starting circuit is open.

CAUTION

This model may be equipped with either 400 or 800 ampere starting fuse depending on starting motor connection. The two fuses are of the same physical size. Observe marking on panel. Do not use an incorrectly rated fuse.

MAIN BATTERY KNIFE SWITCH

This switch is used to connect the batteries to the locomotive low voltage electrical system and should be kept closed at all times during operation.

CAUTION

Do not open battery switch at engine shutdown following load operation. The turbocharger lube oil pump will come on and continue to run for approximately 35 minutes following engine shutdown, then shut off automatically. The 35 minutes allows turbocharger bearings to cool using engine lube oil.

On special order, the turbocharger lube pump is connected on the battery side of the knife switch, so that opening the switch after engine shutdown will not stop the pump. If a fault occurs in the turbocharger lube pump circuit, then a TURBO LUBE PUMP NOT RUNNING message will be displayed.

FUSE TEST EQUIPMENT

To facilitate testing of fuses, a pair of fuse test blocks and a test light are installed on the fuse and switch panel. A test light toggle switch is located on the No. 3 circuit breaker panel. Fuses may be readily tested as follows. Move test light switch to the on position to make sure the fuse test light is not burned out. Move test light switch to the off position to turn light off. Place fuse to be tested across the test blocks so that the metal ends of the fuse are in firm contact with the blocks. If the fuse is good the light will come on.

It is always advisable to test fuses before installation. Always isolate the circuits in question by opening their switches before changing or replacing fuses.

ENGINEROOM EQUIPMENT

Engine starting and monitoring equipment is located in the engineroom as shown in Fig. 2-24.

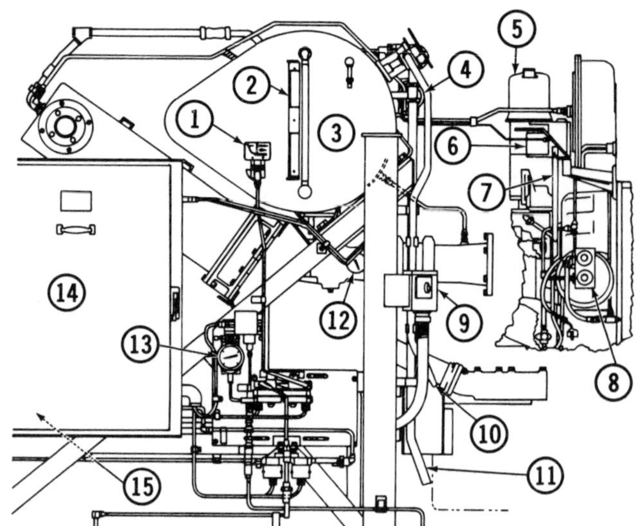

1. Manual shutter Control Valve
2. Water Level Instruction Plate
3. Water Level Sight Glass
4. Lube Oil Pressure Gauge
5. Governor
6. Load Regulator
7. Injector Control Lever (Layshaft)
8. Low Water And Crankcase Pressure Detector (Fig. 2-27)
9. Fuel Prime/Engine Start Switch
10. Water Filler
11. Water Tank Overflow
12. Water Temperature Gauge
13. Air Pressure Gauge
14. AC Cabinet
15. Fuel Oil Filter Bypass Gauge (On other side of equipment rack at the left side of the fuel filters).

Fig. 2-24 Typical Engineroom Equipment

ENGINE STARTING CONTROL

NOTE

Refer to Operation section for complete inspection and starting instructions.

FUEL PRIME / ENGINE START SWITCH

This three position rotary switch, Fig. 2-25, is located in a junction box mounted on the equipment rack. The functions of the three positions are as follows:

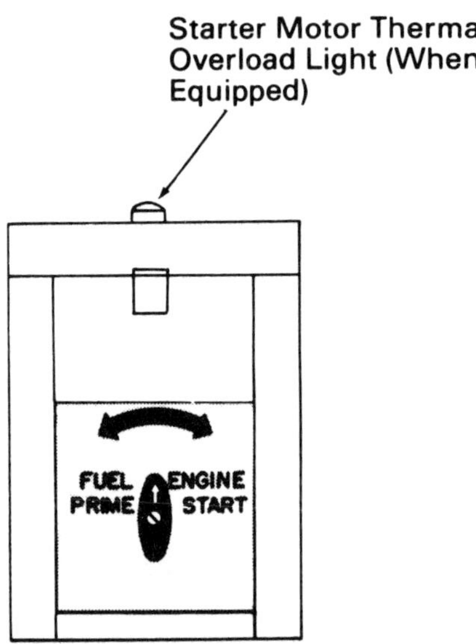

Fig. 2-25 Fuel Prime / Engine Start (FP/ES) Switch

FUEL PRIME Position

This position is used to prime the engine with fuel prior to starting. In this position the fuel pump motor is energized with battery power but the engine will not crank. Additional contacts energize the auxiliary turbocharger lube oil pump motor, ensuring a supply of lube oil under pressure to the turbocharger bearings during startup.

ENGINE START Position

This position is used to supply power from the batteries to the starting motors. The starter motor pinion gear engages with the engine ring gear which causes the engine to crank until PF/ES switch is released.

This position also arms the automatic cooling system drain circuit, if provided, when the WATER DRAIN circuit breaker is closed.

Centered (Off) Position

The FP/ES switch is spring loaded to return to this position when released. Contacts that are normally closed in this position sypply power to the fuel pump motor from the auxiliary generator when the engine is running.

NOTE

When equipped, a light on top of the FP/ES switch junction box will come on to indicate that the starter motors have been overloaded. When this light is on, power will not be applied to the starter motors regardless of FP/ES switch position. The light will go out automatically when starter motors have cooled sufficiently to allow restart attempt.

INJECTOR RACK MANUAL CONTROL LEVER (LAYSHAFT)

This engine mounted hand operated lever, Fig. 2-24, may be used to manually operate the injector racks. It is primarily used to position the injector racks during engine cranking, thereby providing an immediate supply of fuel to the cylinders.

CAUTION

On units equipped with engine purge control system, do not push injector control lever until engine has cranked for 6 seconds.

MONITORING DEVICES

The following devices monitor certain locomotive systems. They provide a visual indication as to the condition of the systems.

Each device represents a system which could cause the engine to shut down. Periodic checks of these systems will alert the operator to an impending failure. Report all abnormal readings to proper maintenance personnel.

WATER LEVEL INSTRUCTION PLATE

This plate is mounted next to a sight gauge on the water tank. To check water level, open round valve handle at bottom of gauge. Read water level using the instruction plate as a guide, then close valve. To avoid false readings drain gauge using small drain cock at bottom of gauge.

LUBE OIL PRESSURE GAUGE

This gauge provides a ready reference indicating lube oil pressure. During normal operation lube oil pressure will increase as diesel engine speed increases.

WATER TEMPERATURE GAUGE

Engine inlet water temperature may be readily checked using this gauge. The gauge is color coded to indicate COLD (blue), NORMAL (green), and HOT (red). Temperature approaching the hot zone may indicate tunnel or similar operation.

FILTER BYPASS GAUGE (If So Equipped)

This gauge indicates the condition of the primary fuel filters. Increased pressure differential across the filters will be indicated by a higher reading on the gauge. As the pressure increases, a bypass valve will begin to open, bypassing the primary fuel filters. This bypassing imposes a filtering burden on the engine mounted fuel oil filters which will shorten their service life.

AIR PRESSURE GAUGE

This gauge indicates No. 1 main air reservoir pressure.

SAFETY DEVICES

A mechanism to detect low engine lubricating oil pressure is built into the engine governor. Under normal operating conditions, engine lubricating oil under pressure is supplied to the mechanism. Should oil pressure drop to a dangerously low level, a small plunger, Fig. 2-26, will pop out the side of the governor body, indicating that the mechanism has tripped. The GOVERNOR SHUTDOWN display message will appear on the screen and the engine will shut down in approximately 2 seconds if operating in throttle positions 4 and above. At idle, and in throttle positions, 1, 2, and 3, a time delay before shutdown is built into the governor.

The locomotive is also equipped with devices which will detect low cooling water pressure, and excessive crankcase pressure, Fig. 2-27, and a hot oil detector.

When activated, the devices release oil pressure from the line leading to the low oil pressure mechanism in the governor, causing engine shutdown.

If necessary to determine cause of shutdown, check the crankcase pressure and low water pressure detecting devices for protruding reset buttons. A protruding button indicates the device that has caused engine shutdown. If crankcase pressure or low water pressure is not the cause, then the engine was shut down by either the hot oil detector or true oil pressure failure.

WARNING

When it is determined that the crankcase pressure detector has tripped, make no further engineroom inspections. Do not attempt to restart the engine. Isolate the unit. Drain the cooling system in accordance with railroad regulations if freezing conditions are possible.

If neither the crankcase pressure nor the low water pressure detector has tripped, and engine oil level is satisfactory with a hot engine condition apparent, do not attempt to restart the engine. Report engine shutdown circumstances to maintenance personnel.

OVERSPEED MECHANISM

An overspeed mechanism is provided to stop injection of fuel into the cylinders should engine speed become excessive. This will resutl in immediate shutdown of the engine and no battery charge / no power alarm.

To reset mechanism, move trip reset lever, Fig. 2-26, counterclockwise until it resets.

MISCELLANEOUS DEVICES

MANUAL SHUTTER CONTROL VALVE

During normal operation this valve, Fig. 2-24, is in the OPERATION position. In this position the cooling control system automatically opens and closes the cooling system shutters, depending on conditions.

In an emergency, the shutters may be opened manually by moving the shutter control valve to the TEST position.

AUTO. DRAIN COLD WATER FILL SWITCH

Operating this pushbutton switch, Fig. 2-24, disarms the automatic cooling system drain circuit. (Pushbutton switch and drain circuit are special equipment.) If automatic drain valve is open, operating the switch causes the valve to close, enabling the cooling system to be refilled.

NOTE

Automatic cooling system drain circuit remains disarmed, that is, it will not protect the cooling system against freezing after cold water fill pushbutton is operated, until fuel prime / engine start switch is turned to the START position.

SPOTTER CIRCUIT (If So Equipped)

The spotter circuit is used to move the locomotive a slight distance. Battery voltage is applied to two traction motor in series when both spotter switches are operated simultaneously. The isolation switch must be in RUN and the alarm bell will ring while the circuit is in operation.

SECTION 3
OPERATION

INTRODUCTION

This section covers recommended procedures for operation of the locomotive. The procedures are briefly outlined and do not contain detailed explanation of equipment location or function.

PREPARATION FOR SERVICE

GROUND INSPECTION

Check for the following:

1. Leakage of fuel oil, lube oil, water or air.
2. Loose or dragging parts.
3. Proper hose connections between units in multiple.
4. Proper positioning of all angle cocks and shut-off valves.
5. Air cut in to truck brake cylinders.
6. Satisfactory condition of brake shoes.
7. Fuel supply.
8. Proper installation of control cables between units.

LEAD UNIT CAB INSPECTION

On the lead or control unit, the control locations described in Section 2 should be checked and the equipment positioned for operation as follows:

FUSE AND SWITCH PANEL

1. Main battery switch closed.
2. All fuses installed and in good condition, and of correct rating as indicated on panel.

CIRCUIT BREAKER PANELS

1. All breakers in the black area of the panels in on position.
2. If automatic cooling system drain circuit (if so equipped) is to protect cooling circuit against freezing after next engine shutdown. WATER DRAIN circuit breaker must be on.
3. Other circuit breakers on as required.
4. At the No. 2 circuit breaker panel, verify that the ground relay cutout switch is closed.

ENGINE CONTROL PANEL

1. Isolation switch in START position.
2. Headlight control switch in proper position for lead unit operation.
3. Dynamic brake cutout switch (if equipped), in DYN. BRAKE (up) position.
4. Miscellaneous switches positioned as required.
5. Remote traction motor cutout switch (if equipped), in MOTORS ALL IN position.

NOTE

The electrical cabinet is pressurized with filtered air. Cabinet doors must be securely closed during locomotive operation.

OPERATOR'S CONTROL STAND

Switches and operating handles on the control stand should be positioned as follows:

1. Place control and fuel pump switch in on (up) position.

2. Place engine run switch and the generator field switches in the off (down) position.
3. If equipped with power reduction circuit, set power reduction toggle switch to OFF position.
4. Light and miscellaneous switches positioned as desired.
5. Move throttle handle to IDLE and dynamic brake handle to OFF position. Position reverser handle to neutral and remove.

26L AIR BRAKE EQUIPMENT

1. Insert automatic brake valve handle (if removed) and place in SUPPRESSION position. This will nullify the application of any safety control equipment used.
2. Insert independent brake valve handle (if removed) and move to FULL APPLICATION position.
3. Position cut-off valve to IN position. On units equipped with a three position cut-off valve, position valve to either FRT or PASS depending on make-up of train.
4. Place multiple unit valve, Fig. 2-11, located at lower left side of control stand in LEAD OR DEAD or in OPEN IN LEAD OR DEAD position as indicated at the valve handle.

STARTING THE DIESEL ENGINE

After the following inspections have been completed, the diesel engine may be started.

ENGINEROOM INSPECTION

The engineroom equipment can be inspected and operated by opening the access doors along the sides of the locomotive long hood.

1. Check air compressor for proper lube oil supply.
2. Check that water level, in water tank sight glass, is near the FULL (ENGINE DEAD) mark on the water level instruction plate.

NOTE

Water level should be rechecked when engine is running. Level should be near FULL (ENGINE RUNNING) mark.

3. Check all valves for proper positioning.
4. Observe for leakage of fuel oil, lube oil, water, or air.

ENGINE INSPECTION

The engine should be inspected before as well as after starting.

1. Check that overspeed mechanism is set.
2. Check that the governor low oil pressure trip plunger is set, and that oil is visible in the governor sight glass.
3. Check that the crankcase (oil pan) pressure and low water pressure detector reset buttons are set (pressed in). If either button protrudes, press and hold button for 5 seconds immediately after engine starts.
4. Check that engine top deck, air box and oil pan inspection covers are in place and are securely closed.
5. Check sight gauge on lube oil filter tank. If gauge is full, proceed to Step 6. If gauge is empty, make certain that oil strainer housing is full. The oil level should be maintained up to the overflow outlet of the housing.
6. Pull out oil level gauge (dipstick) from side of engine oil pan. Oil gauge should be coated with lube oil.

NOTE

A properly filled oil system will coat the oil gauge above the FULL mark when the engine is stopped. To obtain an accurate check, recheck level, when the engine is idling and at normal operating temperature.

ENGINE STARTING

After the preceding inspections have been completed, the diesel engine may be started. Close engineroom doors after engine start.

Perform the following:

NOTE

If engine water temperature is 10°C (50°F) or less, preheat engine before attempting to start. Prelube engine if it has been shut down more than 48 hours. Refer to Engine Maintenance Manual for prelube procedures.

1. Open cylinder test cocks and bar over the engine at least one revolution. Observe for leakage from test cocks. Close test cocks.

   NOTE

   Leakage from cylinder test cocks indicates a problem within the engine. Notify maintenance personnel.

2. Check that all fuses are installed and in good condition, and of the correct rating as indicated on panel. Verify that the main battery and ground relay cutout switches are closed.
3. At the circuit breaker panels, check that all breakers in the black areas are in the on position.
4. At the operator's control stand, make certain that the generator field and engine run switches are off (down). Verify that the control and fuel pump switch is on (up).

   NOTE

   When starting trailing unit diesel engines and control cables have been connected between units, the control and fuel pump switch should remain off.

5. At the engine control panel, verify that the isolation switch is in the START position.
6. At the equipment rack, place the fuel prime / engine start switch in the FUEL PRIME position until fuel flows in the return fuel sight glass, Fig. 3-1, clear and free of bubbles (normally 10 to 15 seconds).

   CAUTION

   On units equipped with engine purge control system, do not push injector control lever until engine has cranked for 6 seconds.

7. Position injector control lever (layshaft) at about one-third rack (about 1.6 on the governor scale), except units equipped with engine purge control system. Move the fuel prime / engine start switch to ENGINE START position. Hold the switch in this position until the engine fires and speed increases, but not more than 20 seconds.

   CAUTION

   Starter motors should not be allowed to crank engine for more than 20 seconds. If engine fails to start after 20 seconds have elapsed, allow 2 minutes for starter cooling.

8. Release injector control lever when engine comes up to idle speed. Do not advance lever to increase speed until oil pressure is confirmed.

   NOTE

   Engine water inlet temperature should be allowed to reach 120°F at idle before load is applied.

9. Check low water pressure detector reset button after engine starts. If tripped, press button to reset detector. The engine will shut down after a short time delay if the detector is not reset.

   NOTE

   If the detector is difficult to reset after engine starts, confirm oil pressure, then position the injector control lever (layshaft) to increase engine speed for a short time, and press the reset button.

10. Check the following with the engine running and at normal operating temperature.
    a. Coolant level is near the FULL (ENGINE RUNNING) mark on the water level instruction plate.

b. Lube oil level is near the FULL mark on oil level gauge (dipstick).

c. Governor oil level.

d. Compressor lube oil level.

TRAILING UNIT CAB INSPECTION

Switches, circuit breakers, and controls located in the cab of a trailing unit should be checked for proper positioning as follows:

FUSE AND SWITCH PANEL

1. Main battery knife switch closed.
2. Fuses installed and in good condition, and of correct rating as indicated on panel.

CIRCUIT BREAKER PANELS

1. All breakers in the black area of the circuit breaker panels in on position.
2. If automatic cooling system drain circuit, special, is to protect cooling system against freezing after next engine shutdown, WATER DRAIN circuit breaker must be on.
3. Other circuit breakers on as required.
4. At the No. 2 circuit breaker panel, verify that the ground relay cutout switch is closed.

ENGINE CONTROL PANEL

1. Isolation switch in START position.
2. Headlight switch in proper position to correspond with unit position in the consist.
3. Dynamic brake cutout switch (if equipped), in DYN. BRAKE position.
4. Miscellaneous switches positioned as required.
5. Remote traction motor cutout switch (if equipped), in MOTORS ALL IN position.

NOTE

The electrical cabinet is pressurized with filtered air. Cabinet doors must be securely closed during locomotive operation.

OPERATOR'S CONTROL STAND

Switches and operating handles on the control stand should be positioned as follows:

1. Control and fuel pump switch, generator field switch, and engine run switch must be off.
2. Move throttle to IDLE and dynamic brake handle to OFF position. Position reverser handle to neutral and remove to lock other handles.
3. If equipped with power reduction circuit, set power reduction toggle switch of OFF position.
4. Light and miscellaneous switches positioned as desired.

26L AIR BRAKE EQUIPMENT

1. Place automatic brake valve handle in HANDLE OFF position. Remove handle (if so equipped).
2. Place independent brake valve handle in RELEASE position. Remove handle (if so equipped).
3. Place cut-off valve to OUT position.
4. Place multiple unit valve in TRAIL or in CLOSED IN TRAIL position as indicated at the valve handle.

STARTING TRAILING UNIT DIESEL ENGINES

Engines in trailing units are started in the same manner as the engine in the lead unit. Refer to "Starting The Diesel Engine" portion of this section.

NOTE

If control jumper cables are already connected between units, ensure that the control and fuel pump, generator field, and engine run switches are off. This will allow these systems to be controlled from the lead unit.

PLACING UNITS ON THE LINE

After the diesel engines are started and inspected, units may be placed on the line as desired by placing the isolation switch on the engine control panel in the cab in the RUN position. If the consist is at a standstill, be certain that the throttle handle in all units is in the IDLE position before placing any unit on the line.

PRECAUTIONS BEFORE MOVING LOCOMOTIVE

The following points should be carefully checked before attempting to move the locomotive under its own power:

1. Make sure that main reservoir air pressure is normal.

   This is very important, since the locomotive is equipped with electro-magnetic switchgear which will function in response to control and permit operation without air pressure for brakes.
2. Check for proper application and release of air brakes.
3. Release hand brake and remove any blocking under the wheels.

> **CAUTION**
>
> It is desirable that engine water temperatures be 49°C (120°F) or higher before full load is applied to the engine. After idling at ambient temperature below −18°C (0°F), increase to full load level should be made gradually.

HANDLING LIGHT LOCOMOTIVE

With the engine started and placed "on-the-line" and the preceding inspections and precautions completed, the locomotive is handled as follows:

1. Place the engine run switch and generator field switch in on (up) position.
2. Place headlight and other lights on as needed.
3. Insert reverser handle and move it to the desired direction of travel, either forward or reverse.
4. Release air brakes.
5. Open throttle to position No. 1, 2, or 3 as needed to move locomotive at desired speed.

   NOTE

   Locomotive response to throttle movement is almost immediate. There is little delay in power buildup.
6. Throttle should be in IDLE before coming to a dead stop.
7. Reverser handle should be moved to change direction of travel only when locomotive is completely stopped.

DRAINING AIR RESERVOIRS AND STRAINERS

The air reservoirs and air strainers or filters should be drained periodically whether or not equipment is provided with automatic drain valves. Follow the maintenance schedule established by the railroad.

COUPLING LOCOMOTIVES TOGETHER

When coupling units together for multiple unit operation, the procedure below should be followed:

1. Couple and stretch units to ensure couplers are locked.
2. Install control cable between units.
3. Attach platform safety chains between units.
4. Perform ground, engineroom, and engine inspections, as outlined in preceding articles.
5. Position cab controls for trailing unit operation as outlined in preceding articles. Remove reverser handles from all controllers to lock controls.

6. Connect air brake hoses between units.
7. Open required air hose cutout cocks on each unit.
8. Make a setup of the brakes on the consist to determine if brakes apply on each unit. Brakes then must be released to determine if all brakes release. The same procedure must be followed to check the independent brake application. Also, release an automatic service application by depressing the independent brake valve handle down. Inspect all brakes in the consist to determine if they are released.

COUPLING UNITS TOGETHER FOR DYNAMIC BRAKING

The locomotive, when equipped with basic dynamic brakes, makes use of electrical potential from the brake control rheostat to control braking strength by controlling excitation of the main generator field. This electrical potential is impressed upon a trainlined wire to control dynamic braking strength of all units in a consist equipped with potential line brake control. However, the total braking effort of a multi-unit consist can become quite high. Carefully observe railroad rules regarding multiple unit dynamic braking in critical service.

COUPLING LOCOMOTIVE TO TRAIN

Locomotive should be coupled to train using the same care taken when coupling units together. After coupling, make the following checks:

1. Test to see that couplers are locked by stretching connection.
2. Connect air brake hoses.
3. Slowly open air valves on locomotive and train to cut in brakes.
4. Pump up air using the following procedure.

PUMPING UP AIR

After cutting in air brakes on train, note the reaction of the main reservoir air gauge. If pressure falls below trainline pressure, pump up air as follows:

1. Place generator field switch in off position.
2. Move reverser handle to neutral position.
3. Open throttle as needed to speed up engine and thus increase air compressor output.

NOTE

Throttle may be advanced to No. 5 if necessary. Engine should not, however, be run unloaded (as in pumping air) at speeds beyond throttle No. 5 position.

BRAKE PIPE LEAKAGE TEST

Prior to operating the 26L brake equipment, a leakage test must be performed. Brake pipe leakage tests should be made in accordance with the railroad operating rules and Power Brake Law.

STARTING A TRAIN

The method to be used in starting a train depends upon many factors such as the type, weight and length of the train and amount of slack in the train; as well as the weather, grade and track conditions. Since all of these factors are variable, specific train starting instructions cannot be provided and it will therefore be up to the operator to use good judgment in properly applying the power to suit requirements. There are, however, certain general considerations that should be observed. They are discussed in the following paragraphs.

A basic characteristic of the diesel locomotive is its high starting tractive effort, which makes it imperative that the air brakes be completely released before any attempt is made to start a train. It is therefore important that sufficient time be allowed after stopping, or otherwise applying brakes, to allow them to be fully released before attempting to start the train.

The locomotive possesses sufficiently high tractive effort to enable it to start most trains without taking slack. The practice of taking slack indiscriminately should thus be avoided. There will, however, be instances in which it is advisable (and sometimes necessary) to take slack in starting a train. Care should be taken in such cases to prevent excessive locomotive acceleration which will cause undue shock.

Proper throttle handling is important when starting trains since it has a direct bearing on the power being applied. As the throttle is advanced, a power increase occurs almost immediately, and power applied is at a value dependent upon throttle position. It is therefore advisable to advance the throttle one notch at a time when starting a train. A train should be started in as low a throttle position as possible, thus keeping the speed of the locomotive at a minimum until all slack has been removed and the train completely stretched. Sometimes it is advisable to reduce the throttle a notch or two at the moment the locomotive begins to move in order to prevent stretching slack too quickly or to avoid slipping.

When ready to start, the following general procedure is recommended.

1. Place isolation switch in RUN position.
2. Move reverser handle to the desired direction, either forward or reverse.
3. Place engine run and generator field switches in the on position.
4. Release both automatic and independent air brakes.
5. Open the throttle one notch every few seconds as follows:
   a. To No. 1 – Loading will stop at a specific low value. This may be noted on the load indicating meter. At an easy starting place the locomotive may start the train.

      NOTE

      The design of the locomotive power control system makes it generally unnecessary to apply locomotive independent brakes or to manipulate the throttle between position No. 1 and IDLE during starting.

   b. To No. 2, 3, or higher (experience and the demands of the schedule will determine this) until the locomotive moves.
6. Reduce throttle one or more notches if acceleration is too rapid.
7. After the train is stretched, advance throttle as desired.

NOTE

The wheel control system, operating in Super Series, reacts automatically to maximize tractive effort for any particular operating condition. These conditions include wet rails, oiled rails, and full throttle operation for maximum acceleration or climbing hills. The wheel control system is totally automatic and requires no operator action. Therefore the wheel slip light will never come on while in Super Series, above 1.5 mph, unless a wheel overspeed, locked powered wheel, or Super Series failure condition exists.

ACCELERATING A TRAIN

After the train has been started, the throttle can be advanced as rapidly as desired to accelerate the train. The speed with which the throttle is advance depends upon demands of the schedule and the type of locomotive and train involved. In general, however, advancing the throttle one notch at a time is desired to prevent slipping. The load indicating meter provides the best guide for throttle handling when accelerating a train. By observing this meter it will be noted that the pointer moves toward the right (increased amperage) as the throttle is advanced. As soon as the increased power is absorbed, the meter pointer begins moving toward the left. At that time, the throttle may again be advanced. Thus for maximum acceleration without slipping, the throttle should be advanced one notch each time the meter pointer begins moving toward the left until full power is reached in throttle position No. 8.

A short time operation nameplate is located directly beneath the load current indicating meter.

AIR BRAKING WITH POWER

The method of handling the air brake equipment is left to the discretion of the individual railroad. However, when braking with power, it must be remembered that for any given throttle position, the draw bar pull rapidly increases as the train speed decreases. This pull might become great enough to part the train unless the throttle is reduced as the train speed decreases. Since the pull of the locomotive is indicated by the amperage on the load meter, the operator can maintain a constant pull on the train during a slow down by keeping a steady amperage on the load meter. This is accomplished by reducing the throttle a notch whenever the amperage starts to increase. It is recommended that the independent brakes be kept fully released during power braking. The throttle must be in IDLE before the locomotive comes to a stop.

OPERATING OVER RAIL CROSSING

When operating the locomotive at speeds exceeding 25 mph, reduce the throttle to No. 4 position at least eight seconds before the locomotive reaches a rail crossing. If the locomotive is operating in No. 4 position or lower, or running less than 25 mph, allow the same interval and place the throttle in the next lower position. Advance the throttle after all units of the consist have passed over the crossing. This procedure is necessary to ensure decay of motor and generator voltage to a safe level before the mechanical shock that occurs at rail crossings is transmitted to the motor brushes.

RUNNING THROUGH WATER

Under absolutely no circumstances should the locomotive be operated through water deep enough to touch the bottom of the traction motors. Water any deeper than 3″ above the rail is likely to cause traction motor damage.

When passing through any water on the rails, exercise every precaution under such circumstances and always go very slowly, never exceeding 2 to 3 mph.

WHEEL CONTROL SYSTEM

The starting wheel slip system consists of electronic circuits separate from the Super Series wheel control system. The starting wheel slip system provides a backup for the Super Series system in the event of Super Series failure. If all conditions are present for Super Series operation, the locomotive will automatically switch into Super Series operation.

The starting wheel slip system is a modified rate-of-change corrective type system. It reacts to the severity of the wheel slip by smoothly reducing power and applying sand. Depending upon the severity of the slipping condition, the wheel slip light may or may not flash on and off as the slip is corrected.

The Super Series system is not a corrective type wheel slip system. It is a wheel control system which allows each traction motor to seek maximum tractive effort. In some cases this involves allowing the wheels to creep (rotate at a slightly faster than ground speed) while being controlled. Under severe rail conditions sand may be applied automatically along with smooth power control. The WHEEL SLIP light will come on during wheel control while the locomotive is in Super Series.

LOCOMOTIVE SPEED LIMIT

The maximum speed at which the locomotive can be safely operated is determined by the gear ratio. This ratio is expressed as a double number such as 70:17. The 70 indicates the number of teeth on the axle gear while the 17 represents the number of teeth on the traction motor pinion gear.

Since the two gears are meshed together, it can be seen that for this particular ratio the motor armature turns approximately four times for a single revolution of the driving wheels. The locomotive speed limit is therefore determined by the maximum permissible rotation speed of the motor armature. Exceeding this maximum could result in serious damage to the traction motors.

Although not basically applied, overspeed protective equipment is available for installation on locomotives. The equipment consists of an electro-pneumatic arrangement with many possible variations to suit specific requirements. In general, however, an electrical switch in the speed recorder is used to detect the overspeed. This switch in turn initiates certain air brake functions which reduce the train speed.

MIXED GEAR RATIO OPERATION

If the units of the consist are of different gear ratios, the locomotive should not be operated at speeds in excess of that recommended for the unit having the lowest maximum permissible speed. Similarly, operation should never be slower than the minimum continuous speed (or maximum motor amperage) for units having established short time ratings.

To obtain a maximum tonnage rating for any single application, Electro-Motive will, upon request, analyze the actual operation and make specific tonnage rating recommendations.

DYNAMIC BRAKING

> **WARNING**
> The dynamic brake system is disconnected when operating with a traction cut out.

Dynamic braking, on locomotives so equipped, can prove extremely valuable in retarding train speed in many phases of locomotive operation. It is particularly valuable while descending grades, thus reducing the necessity for using air brakes.

Maximum braking strength is obtained at a specific speed, depending on locomotive gear ratio. Basic locomotives equipped with 70:17 gear ratio, obtain maximum braking strength at approximately 29 mph. At train speeds higher than the optimum, braking effectiveness gradually declines as speed increases. For this reason, it is important that dynamic braking be started BEFORE train speed becomes excessive. While in dynamic braking, the speed of the train should not be allowed to "creep" up by careless handling of the brake.

To operate dynamic brakes, proceed as follows:

1. The reverser handle must be positioned in the direction of the locomotive movement.
2. Return throttle to idle and hold it in idle for 10 seconds before proceeding.

> **WARNING**
> The 10 seconds delay must be accomplished before the braking handle is moved into SET UP position. This delay occurs automatically when the throttle is placed in idle.
>
> It is possible for a sudden surge of braking effort to occur if the dynamic braking handle is open when the automatic delay times out.

3. Move the braking handle into SET UP position. This establishes the dynamic braking circuits. It will also be noted that a slight amount of braking effect occurs, as evidenced by the load current indicating meter.
4. After the slack is bunched, the dynamic braking handle is moved to control dynamic braking strength.

NOTE

When the dynamic brake handle is advanced out of SET UP, the control system sets engine speed to ensure sufficient cooling air for the traction motors.

5. Braking effort may be increased by slowly advancing the handle to FULL 8 position if desired. Maximum braking current can occur over a wide range of braking handle positions. This range allows braking effort to increase as train speed increases. The tendency is to hold train speed relatively constant for a given braking handle position when conditions result in less than the maximum allowable current.

NOTE

On units equipped for "Grid Current Trainline Control" of dynamic braking, current is limited by dynamic brake handle position, with maximum grid current obtainable only when the brake handle is in the FULL 8 position. Braking current will generally be at or near the maximum obtainable at the given handle position, for a given handle position is not as effective as with the basic dynamic brake.

6. With automatic regulation of maximum braking strength, the brake warning light on the controller should seldom give indication of excessive braking current. If the brake warning light does flash on however, stop advancing the braking handle until the light goes out.

7. If the light fails to go out after several seconds, move the braking handle back slowly until the light does go out. After the light goes out, the handle may again be advanced to increase braking effort.

NOTE

The brake warning light circuit is "trainlined" so that a warning will be given in the lead unit if any unit in the consist is generating excessive current in dynamic braking. Thus regardless of the load indicating meter reading or braking handle position (which may be less than maximum), whenever the warning light comes on, it should not be allowed to remain on for any longer than two or three seconds before steps are taken to reduce braking strength.

If brake warning indications are repeated, the locomotive should be taken out of dynamic braking and the dynamic brake cutout switch on the engine control panel of the affected unit should be placed in CUTOUT position. The locomotive consist will then operate normally under power and during dynamic braking, but with reduced total braking effort.

8. When necessary, the automatic brake may be used in conjunction with the dynamic brake. However, the independent brake must be KEPT FULLY RELEASED whenever the dynamic brake is in use, or the wheels may slide. As the speed decreases the basic dynamic brake becomes less effective. When the speed further decreases, it is permissible to completely release the dynamic brake by placing the handle in OFF position, applying the independent brake simultaneously to prevent the slack from running out.

CAUTION

When operating in dynamic brake, an air brake application caused by an emergency or safety control condition will disconnect the dynamic brake system.

The locomotive can be operated in dynamic braking when coupled to older units that are not equipped with brake current limiting regulators. If all the units are of the same gear ratio, the unit having the lowest maximum brake current rating should be placed as the lead unit in the consist. The operator can then operate and control the braking effort up to the limit of the unit having the lowest brake current rating, without overloading the dynamic brake system of a trailing unit. The locomotive consist MUST always be operated so as not to exceed the braking current of the unit having the lowest maximum brake current rating.

Units equipped with dynamic brake current limiting regulators can be operated in multiple with other locomotives in dynamic braking regardless of the gear ratio or difference in the maximum brake current ratings.

DYNAMIC BRAKE WHEEL SLIP CONTROL

During dynamic braking, each series group of two traction motors is connected in parallel with each dynamic braking resistor grid circuit and with the other series connectted traction motors. With this arrangement, when a wheel slips it may be motored by other motors in the system. This in effect makes a wheel slip during dynamic braking somewhat self correcting. However, the parallel arrangement of dynamic braking resistor grids and traction motors is such that the full response of the wheel slip control system is available during braking as well as during power operation. The precise and immediate regulation maintained, plus the motoring effect created by the parallel arrangement, provides extremely stable dynamic brake operation.

In addition to the above, a circuit is employed to protect against the possibility of simultaneous slips that otherwise may not be detected.

When a pair of wheels is detected tending to rotate at a slower speed, the retarding effort of the traction motors in the unit affected is reduced (traction alternator field excitation is reduced in the unit affected) and sand is automatically applied to the rails. When the retarding effort of the traction motors in the unit is reduced, the tendency of the wheel set to rotate at a slower speed is overcome. After the wheel set resumes normal rotation, the retarding effort of the traction motors returns (increases) to its former value. Automatic sanding continues for 3 to 5 seconds after the wheel slide tendency is corrected.

OPERATION IN HELPER SERVICE

Basically, there is no difference in the instructions for operating the locomotive as a helper or with a helper. In most instances it is desirable to get over a grade in the shortest possible time. Thus, wherever possible, operation on the grades should be in the full throttle position. The throttle can be reduced, however, where wheel slips cause lurching that may threaten to break the train.

ISOLATING A UNIT

When it becomes necessary to isolate a locomotive unit, observe the following:

1. When operating under power in a multiple unit consist, a unit may be isolated at any time, but discretion as to timing and necessity should be used.
2. When operating in dynamic braking, it is important to get out of dynamic braking before attempting to isolate the unit. This is done by reducing the braking handle to OFF. The isolation switch can then be moved to ISOLATE position to eliminate the braking on that unit. If the braking is resumed, other units will function normally.

CHANGING OPERATING ENDS

When the locomotive consist includes two or more units with operating controls, the following procedure is recommended in changing from one operating end to the opposite end on locomotives equipped with 26L brakes.

ON END BEING CUT OUT

1. Move the automatic brake valve handle to service position and make a 20-pound reduction.
2. After brake pipe exhaust stops, place cut-off valve in OUT position by pushing knob in and turning to the desired position.
3. Place independent brake handle in fully released position.
4. Place multiple unit valve in TRAIL or in CLOSED IN TRAIL position as indicated at the valve handle.
5. Position the automatic brake valve handle in the handle off position.
6. With dynamic brake handle in OFF position and throttle in IDLE, place the reverser handle in neutral position and remove to lock the controls.
7. Place all switches in the off position. Be absolutely certain that the control and fuel pump switch, generator field switch, and engine run switch are in the off position.
8. At the engine control panel, place headlight control switch in proper position for trailing unit operation. Place other switches on as needed.
9. At the circuit breaker panels, all circuit breakers in the black areas are to remain in the on position.
10. After completing the operations outlined in the preceding steps, move to the cab of the new lead unit.

ON END BEING CUT IN

1. At the control stand, make certain the generator field switch is off.
2. Insert reverser handle and leave in neutral position.
3. Place automatic brake valve handle in suppression position to nullify any safety control, overspeed, or train control used.
4. Insert independent brake valve handle (if removed) and move handle to full independent application position.
5. Position cut-off valve to IN position. On units equipped with a three position cut-off valve, position valve to either FRT or PASS depending on make-up of train.

6. Place dual ported cutout cock in OPEN IN LEAD or DEAD position.
7. At the circuit breaker panels, check that all circuit breakers in the black areas are in the on position.
8. At the engine control panel, place the headlight control switch in proper position, and other switches on as needed.
9. Place the engine run, control and fuel pump, and generator field switch in on position. Other switches may be placed on as needed.

STOPPING ENGINE

There are six ways to stop the engine:

1. Press stop button on engine control panel.
2. Press emergency fuel cut-off button.

   Emergency fuel cut-off pushbuttons are located near each fuel filter opening. These pushbuttons operate in the same manner as the stop button and need not be held in nor reset.
3. Use injector control lever.

   The injector control lever can be operated to override the engine governor and move the injector racks to the no fuel position.
4. Close the low water detector test cock.

   When the low water detector trips, oil is dumped from the governor low oil shutdown device, stopping the engine.
5. Use throttle handle.

   To stop all engines "on-the-line" in a consist simultaneously from the cab of the lead unit, move the throttle to the IDLE position, pull the handle out and away from the controller, and move it beyond IDLE to the STOP position.
6. Pull out low oil shutdown plunger on the side of the governor.

FREEZING WEATHER PRECAUTIONS

As long as the diesel engine is running, the cooling system will be kept adequately warm regardless of ambient (outside) temperatures encountered. If is only when the engine is shut down or stops for any reason that the cooling system requires protection against freezing.

Whenever the engine is shut down, and freezing temperatures are possible, the cooling system, as well as the flush toilet and the water cooler, if so equipped, should be drained or otherwise protected from freezing. The automatic cooling system drain circuit, if provided will protect the cooling system.

DRAINING THE COOLING SYSTEM (Basic)

When necessary to drain system, open engine water drain valve located at the pit between the engine and accessory rack. This valve will drain the engine, water tank, water cooled air compressor and associated piping.

CAUTION
If a hot engine is drained, always allow the engine to cool before refilling with coolant.

AUTOMATIC COOLING SYSTEM DRAIN CIRCUIT (If So Equipped)

If the circuit is armed, it will activate the automatic drain valve in freezing temperatures to protect the engine and air compressor cooling systems. If the circuit is *not* armed, and both the battery knife switch and the CONTROL circuit breakers are closed, the WATER DRAIN ACTIVATED OR BREAKER DOWN message will be displayed and the alarm bell will be sounding. To manually drain the system, use the manual drain valve, not the test switch on the automatic valve. Both valves are located in the pit between the engine and the accessory rack.

To arm the circuit for automatic operation, proceed as follows.

1. Engine should be stopped. If you do not want the engine to start during this procedure, open the battery knife switch.
2. Close the WATER DRAIN circuit breaker.
3. Turn the fuel prime / engine start switch to the ENGINE START position for a moment. Circuit should be armed.
4. Open and close the automatic drain valve by momentarily operating the test switch on the valve. Valve operation indicates successful circuit arming. Repeat Step 3 if valve does not operate.
5. Open the battery knife switch and CONTROL circuit breaker unless they should remain closed for some other reason.

   The WATER DRAIN circuit breaker must remain closed and the AUTO. DRAIN COLD WATER FILL switch on the AC cabinet must not be operated as long as the circuit is to remain armed.

To override the automatic cooling system drain circuit and close the automatic drain valve, press the AUTO. DRAIN COLD WATER FILL pushbutton switch on the side of the AC cabined. Alarm bell will sound and WATER DRAIN ACTIVATED OR BREAKER DOWN message will be displayed.

CAUTION

Allow a hot engine to cool down before refilling the cooling system.

After completing refill of cooling system and warming up engine, re-arm the circuit by following the previous procedure.

DRAIN FLUSH TOILET (If So Equipped)

1. Flush toilet until all water has drained from tank.
2. Turn off electric toilet tank heater (if so equipped).
3. Remove pipe plug from bottom of toilet flush piping.

DRAIN WATER COOLER (If So Equipped)

1. Remove and empty water bottle.
2. Drain remaining water in cooler by holding in the spigot button.
3. Turn off electric power to water cooler (if so equipped).

TOWING LOCOMOTIVE IN TRAIN

When a locomotive unit equipped with 26L air brakes is place within a train consist to be towed, control and air brake equipment should be set as follows:

1. Drain all air from main reservoirs and air brake equipment unless engine is to remain idling.
2. Place the multiple unit valve in LEAD OR DEAD or in OPEN IN LEAD OR DEAD position as indicated at the valve handle.
3. Place cut-off valve in OUT position.
4. Place independent brake valve handle in release position.
5. Place automatic brake valve handle in handle off position.
6. Cut in dead engine feature by turning cutout cock, Fig. 2-8, to open (90° to pipe) position. Dead engine cock is located beneath cab floor and may be reached through an access door of locomotive.
7. If engine is to remain idling, switches should be positioned as follows:
   a. Isolation switch in START position.
   b. Battery switch and ground relay cutout switch closed.

c. Generator field circuit breaker OFF.

d. All breakers in black areas of circuit breaker panels in ON position.

e. Starting fuse should be removed. Other fuses should be left in place.

f. Control an fuel pump switch on (up).

g. Fuel pump circuit breaker ON.

h. Throttle in IDLE, dynamic brake handle in OFF position. Remove reverser handle from controller to lock the controls.

8. If a locomotive is to be towed dead in a consist, switches should be positioned as follows:

   a. Battery switch open.

   b. All circuit breakers OFF.

   c. All control switches OFF.

   d. Starting fuse removed.

   e. Throttle in IDLE, dynamic brake handle in OFF position. Remove reverser from controller to lock the controls.

NOTE

If there is danger of freezing, the engine cooling system should be drained. Refer to Freezing Weather Precautions.

LEAVING LOCOMOTIVE UNATTENDED

If at any time it is necessary to leave the locomotive unattended while the engine is running, the following procedure should be adhered to.

1. Observe all railroad safety precautions.
2. Place engine run and generator field switches in the off (down) position.
3. Place throttle in IDLE and dynamic brake handle in OFF position. Remove reverser handle from controller to lock the controls.

SECTION 4
INDICATOR LIGHT AND COMPUTER DISPLAY MESSAGES

This section contains information pertaining to locomotive indicator lights and computer display messages. The section is divided into two parts with Computer Display Messages preceding Indicator Light Messages. The Computer Display Messages section is divided into five subsections:

1. Diesel Engine
2. Traction Power
3. Dynamic Brake
4. Mechanical Faults
5. Accessories and Special Equipment

Each message is described in terms of the overall effect on the locomotive, the condition that produced the effect, and if applicable, the response of the crew.

NOTE

All of the messages listed may not be available on a particular locomotive order. The letter "N" appearing in this section indicates a number. The computer display message will show the actual number.

COMPUTER DISPLAY MESSAGES	EFFECT ON LOCOMOTIVE	CONDITIONS	ACTION BY CREW
DIESEL ENGINE			
ENG AIR FILT DIRTY-THROTTLE 6 LIMIT	Reduced power.	Filters plugged.	No action required.
FUEL PUMP NOT RUNNING [+ alarm]	Engine stops.	Engine running and fuel pump feedback is lost due to a failure in the fuel pump circuit.	Check if fuel pump motor is operating. If motor is operating, then isolate the unit. If motor is not operating, then check fuel pump circuit breaker.
GOVERNOR SHUTDOWN [+ alarm]	Engine stopped.	Low water or crankcase pressure or hot oil or low oil pressure.	Reset LOS.
HOT ENGINE - THROTTLE 6 LIMIT [+ alarm]	Reduced power.	Temporary operating condition-tunnel, desert, etc.	No action required unless message persists. If message persists more than a few minutes, then check radiator blowers, shutters, and coolant level. If coolant level is low or leaking, then shut down the unit. If shutters are closed, make sure that manual shutter control valve is set to OPERATE position. If shutters do not open, then shut down the unit.
START MOTOR OVERLOAD [+ alarm]	Engine will not crank.	Starting motors operated for 20 seconds.	No action required-automatic 2 minute time delay before starting motors will operate.
TURBO CIRCUIT BREAKER DOWN	Engine will not crank.	Turbo lube pump motor circuit breaker is down.	Reset circuit breaker.
TURBO LUBE PUMP NOT RUNNING [+ alarm]	Engine will not crank.	Turbo pump failed to operate prior to engine start.	Isolate unit.
		Turbo pump failed to operate after engine shutdown.	Isolate unit – immediately restart engine and run at idle for 45 minutes.
TRACTION POWER			
POWER-OVEREXCITATION	Cycling of GFC causes rough power regulation.	Main generator field current too high.	No action required.

COMPUTER DISPLAY MESSAGES	EFFECT ON LOCOMOTIVE	CONDITIONS	ACTION BY CREW
LOW HORSEPOWER	Reduced Power.	Main generator output power is below 70% of power demand for 20 minutes.	No action required.
NO SUPER SERIES-REGULATION FAILURE	Transfers to backup wheelslip system.		No action required.
NO SUPER SERIES-UNDERLOADING	Transfers to backup wheelslip system.		No action required.
OPEN TRACTION MOTOR CIRCUIT #N (N = 1, 2, 3, 4, 5, 6) [+ alarm]	Reduced power	One traction motor feedback current is zero.	Cut out motor.
NO LOADING-GROUND RELAY CUT-OUT	Unit will not load.	Ground relay cutout switch open.	Close ground relay cutout switch.
NO POWER-GROUND RELAY LOCK OUT	Unit will not load.	Excessive number of ground relay pickups within a time period while in power operation.	A ground relay lockout reset switch is located on the No. 2 circuit breaker panel of the electrical cabinet. Refer to railroad regulations before resetting the lockout function.
GROUND RELAY-POWER [+ alarm] ↓ ↓ ↓ ↓ TRACTION MOTOR N FLASHOVER OR	Unit will not load.	High voltage ground (flashover).	Automatic reset for first 3 ground relay pickups, then locks out. Cutout motor (if so equipped) or, if allowed by railroad regulations, operate ground relay lockout reset switch on No. 2 circuit breaker panel of the electrical cabinet.
VOLTAGE LIMITING DUE TO GROUND RELAY	Reduced power.	Ground fault	Automatic ground relay reset and voltage limiting. Continued ground faults result in lockout. On units not equipped with motor cutouts, the ground relay is locked out on the fourth pick-up within 10 minutes. On units equipped with motor cutouts, three ground relay pick-ups are allowed at each motor cutout position before lockout occurs. Isolating the unit will generally restore main generator operating voltage to maximum level. If a lockout occurs, then locomotive operation can be resumed by pushing the ground relay lockout reset switch on the No. 2 circuit breaker panel of the electrical cabinet. Refer to railroad regulations before resetting the lockout function.
DYNAMIC BRAKES			
NO DYNAMIC BRAKE-DYNAMIC BRAKE CUT OUT	No dynamic brake on this unit. (Not trainlined)	Dynamic brake cutout switch set in CUTOUT position.	Remember that unit won't perform dynamic braking.

COMPUTER DISPLAY MESSAGES	EFFECT ON LOCOMOTIVE	CONDITIONS	ACTION BY CREW
GROUND RELAY-BRAKE [+ alarm]	Unit will not load.	High voltage ground.	The locomotive is equipped with an automatic ground relay reset function. This computer function will automatically reset the ground relay a certain number of times in a time period before it locks out. The exact lockout conditions depend on railroad specifications. A ground relay lockout reset switch is located on the No. 2 circuit breaker panel of the electrical cabinet. Refer to railroad regulation before resetting the lockout function.
NO DYNAMIC BRAKE-GROUND RELAY LOCK OUT	Unit will not load.	Excessive number of ground relay pickups within a time period while in dynamic brake operation.	A ground relay lockout reset switch is located on the No. 2 circuit breaker panel of the electrical cabinet. Refer to railroad regulations before resetting the lockout function.
NO DYNAMIC BRAKE-GRID BLOWER OPEN	No dynamic brake.	Grid blower failed.	As per railroad policy cut out dynamic brake with cutout switch on engine control panel.
NO DYNAMIC BRAKE-GRID BLOWER SHORTED	No dynamic brake.	Grid blower failed.	As per railroad policy cut out dynamic brake with cutout switch on engine control panel.
NO DYNAMIC BRAKE-GRID OPEN	No dynamic brake.	Grid open.	As per railroad policy cut out dynamic brake with cutout switch on engine control panel.
MECHANICAL FAULTS			
LOCKED POWERED WHEEL	Unload.	Wheel locked with power applied.	Correct problem.
AIR COMPRESSOR-LOW OIL PRESSURE	Engine stops.	Low oil pressure in air compressor.	Isolate unit.
#N AXLE LOCKED (N = 1, 2, 3, 4, 5, 6) [+ alarm]	Unload	Axle locked.	1. Stop train. 2. Look for unit with locked wheel indication. 3. Roll train slowly and observe wheels. a. If wheel slides, then cut unit out of train. b. If all wheels rool and locked wheel indication is no longer visible, then proceed normally.
SLIPPED PINION-TRACTION MOTOR #N (N = 1, 2, 3, 4, 5, 6) [+ alarm]	Unload.		Isolate unit.
ACCESSORIES AND SPECIAL EQUIPMENT			
ARC MODULE FAILED OR MISSING	No effect.	Failure in arc module or associated circuit.	Report to maintenance personnel.
NO LOADING-NO AUX GEN OUTPUT	No power.	Engine running and no main gen. output, or	Check AUX. GEN. and AUX. GEN. FIELD circuit breakers.

COMPUTER DISPLAY MESSAGES	EFFECT ON LOCOMOTIVE	CONDITIONS	ACTION BY CREW
		D14/D18 output, or aux. gen. output.	If a breaker is not tripped, then isolate the unit. If a breaker is tripped, then reset the breaker. If another trip occurs, then isolate the unit.
ENGINE SPEED INCREASE-LOW AIR PRESS.	Increased idle speed.	Low main reservoir pressure. [On some special equipped units, engine speedup due to low main reservoir pressure will occur only when the reverser is put in forward or reverse.]	No action required.
ENGINE SPEED INCREASE-LOW WATER TEMP.	Increase idle speed.	Low water temperature.	No action required.
ENGINE SPEED INCREASE-TRACTION MOTOR COOLING.	Increase idle speed.	Insufficient traction motor cooling in dynamic brake.	No action required.
FAULT MESSAGE STORED.	No effect.	A fault message has been put in archive. **NOTE** "FAULT MESSAGE STORED" will only be displayed when unit is isolated and throttle is in idle.	No action required.
FILTER BLOWER MOTOR CIRCUIT BREAKER DOWN	Improper air filtration.	Inertial filter blower motor circuit breaker is down.	Reset breaker.
NO LOADING-NO D18 OUTPUT	No power.	Engine running and no D14/D18 output but output at aux. gen.	Check A.C. CONTROL circuit breaker. If the breaker is not tripped, then isolate the unit. If the breaker is tripped, then reset the breaker. If another trip occurs, then isolate the unit.
NO SUPER SERIES-CHECK RADAR	Transfers to backup wheelslip system.		No action required.
WATER DRAIN ACTIVATED OR BREAKER DOWN [+ alarm]	No automatic protection of cooling system in low ambient temperatures.	Automatic water drain system inoperative.	Check water drain circuit breaker for tripped condition. If AUTO. DRAIN COLD WATER FILL switch on A.C. cabinet was operated, then fuel prime / engine start switch must be moved to START position to reset the water drain system.
7TH THROTTLE KNOCKDOWN	Throttle 7 operation in throttle 8 position.	Lead unit engine speed reduced to throttle 7.	Check 7th Throttle Knockdown switch on control stand (if so equipped).
WHEEL SLIP Light [+ alarm] ↓ ↓ ↓ ↓			
Intermittent Speed under 1 1/2 mph)		Normal wheel slip correction under severe conditions.	No action required. Do not reduce throttle unless slipping threatens to break the train.

COMPUTER DISPLAY MESSAGES	EFFECT ON LOCOMOTIVE	CONDITIONS	ACTION BY CREW
Continuous (or regular flashing)		Locked sliding wheels.	Check that all locomotive wheels rotate freely. Do not operate locomotive unless all wheels rotate freely.
Irregular flashing (speed over 1 1/2 mph)		Possible Super Series wheel creep system failure.	No action required.
PCS OPEN Light	Penalty brake application and loss of power.	Emergency air brake application.	Move throttle to idle. Move automatic brake to emergency and wait 45 seconds, then to release. Operate P.C. RESET switch (if so equipped) on control stand.
	Emergency brake application and loss of power.	Emergency air brake application.	Move throttle to idle. Move automatic brake to emergency and wait 45 seconds, then to release. Operate P.C. RESET switch (if so equipped) on control stand.
BRAKE WARN Light		Excessive dynamic brake current.	Reduce dynamic brake handle position immediately. If light stays on, then move dynamic brake cutout switch on engine control panel to cutout position.
OSC HDLT Light	Red oscillating headlight is on.	Emergency or penalty brake application or signal light control switch on control stand is in RED position.	
SAND LIGHT	Sanding when below 5 mph.	One of the manual sanding switches is activated.	
S.C. NOT OPER. Light	Automatic train speed system is not operating.	Mode selector switch on the cab indicator is OFF or the speed control switch on the controller is OFF.	

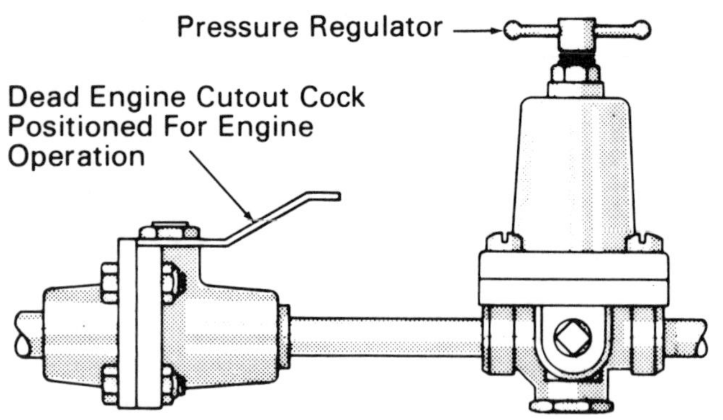

Fig 2-8 Dead Engine Cutout

Cock and Pressure Regulator

Dual Ported
Cutout Cock

MU-2A Valve

Fig. 2-11 Multiple Unit Valve

Fig. 2-13 Typical Indicator Light Panel

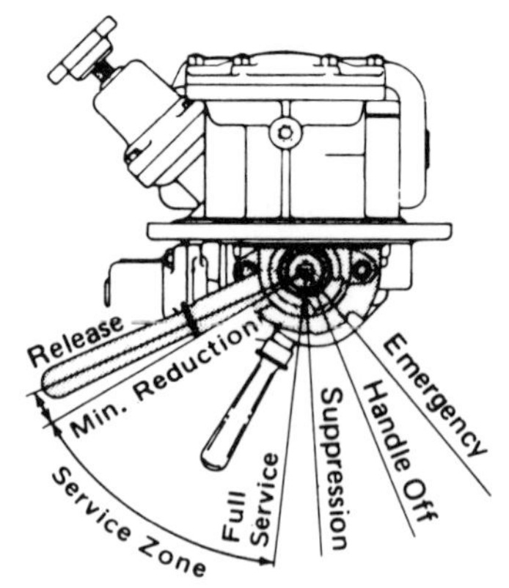

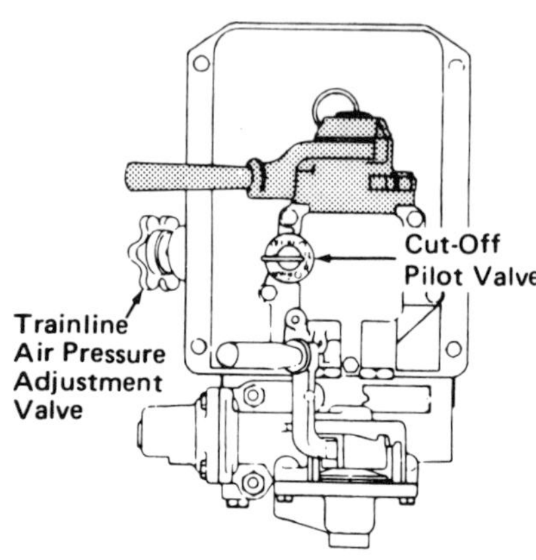

Fig. 2-9 Automatic Brake Valve

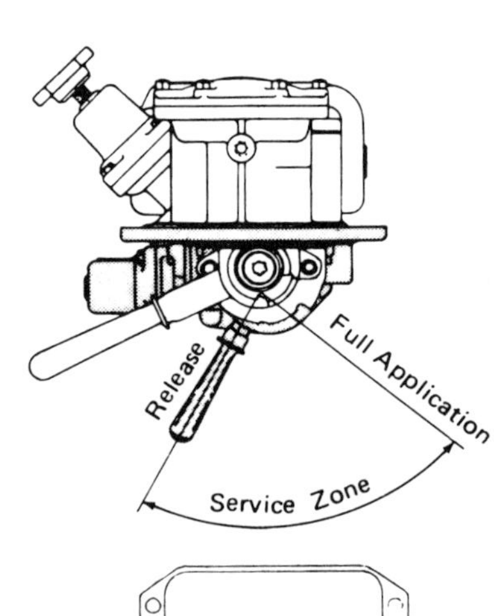

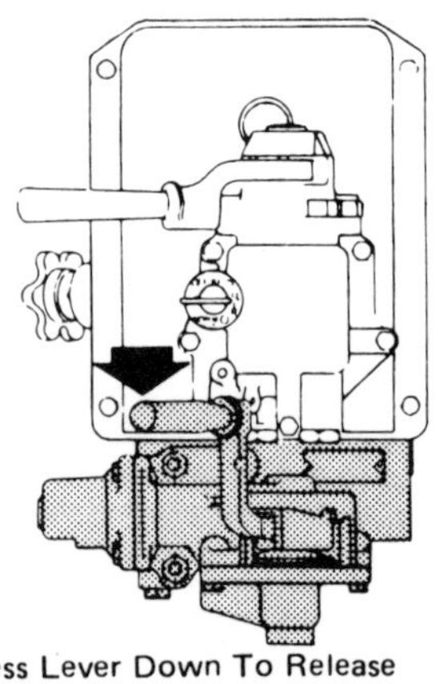

Fig. 2-10 Independent Brake Valve

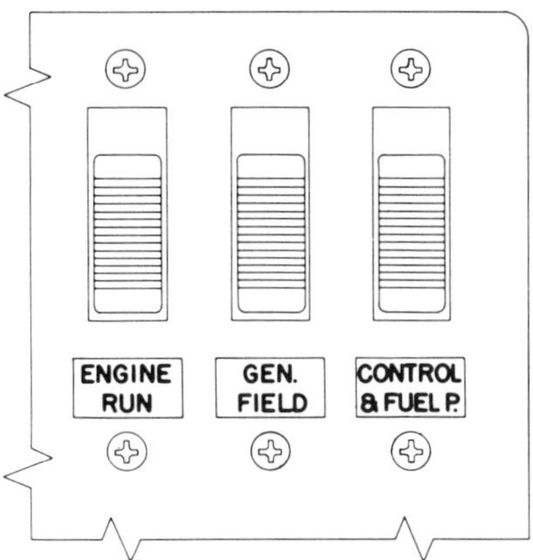

Fig. 2-14 Control and Operating Switches

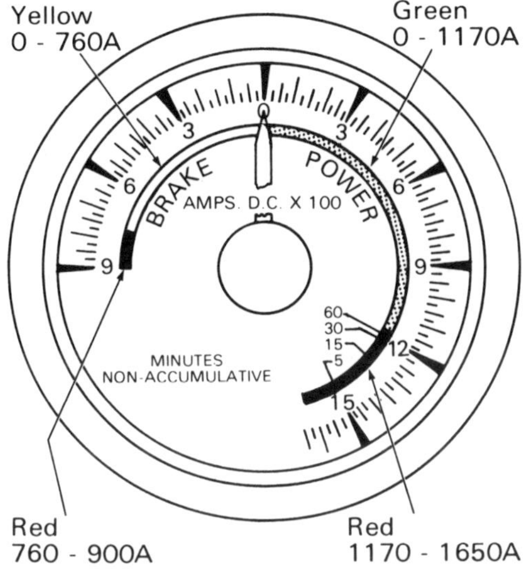

Fig. 2-15 Load Current Indicating Meter

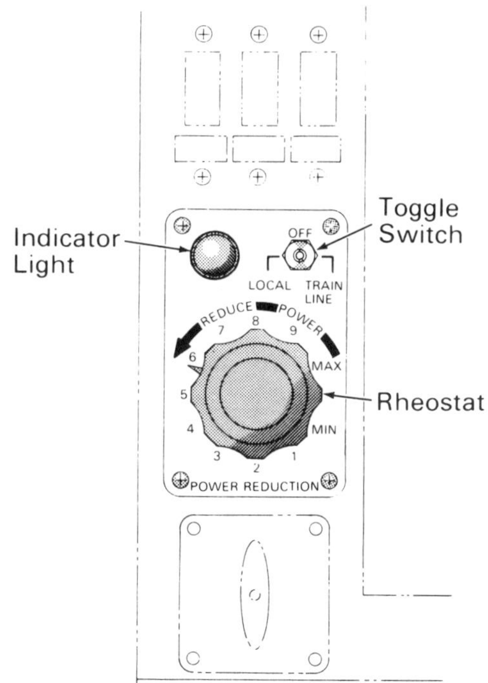

Fig. 2-16 Power Reduction Controls (If Provided)

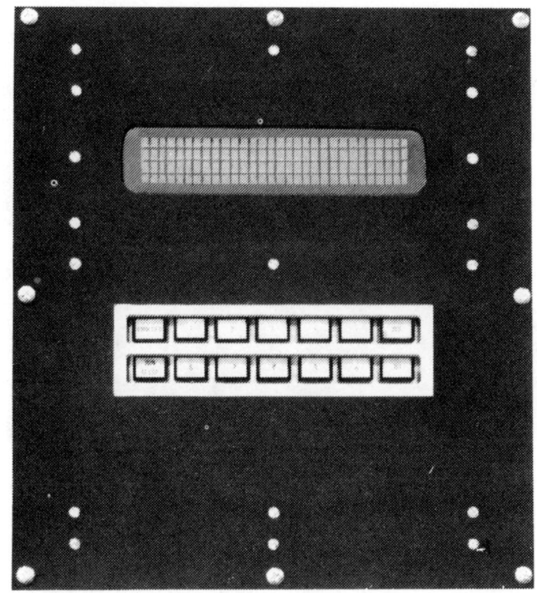

Fig. 2-18 Computer Display / Diagnostic Panel

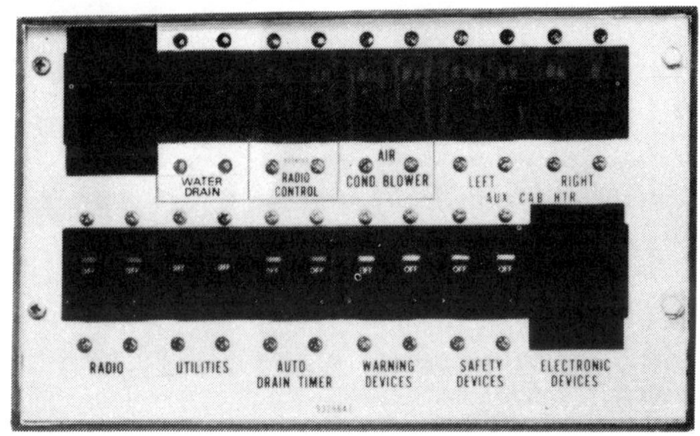

Fig. 2-20 Typical No. 1 Circuit Breaker Panel

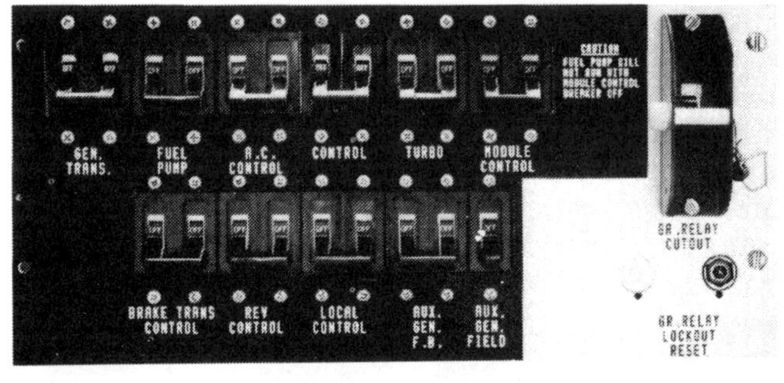

Fig. 2-21 Typical No. 2 Circuit Breaker Panel

Fig. 2-22 Typical No. 3 Circuit Breaker Panel

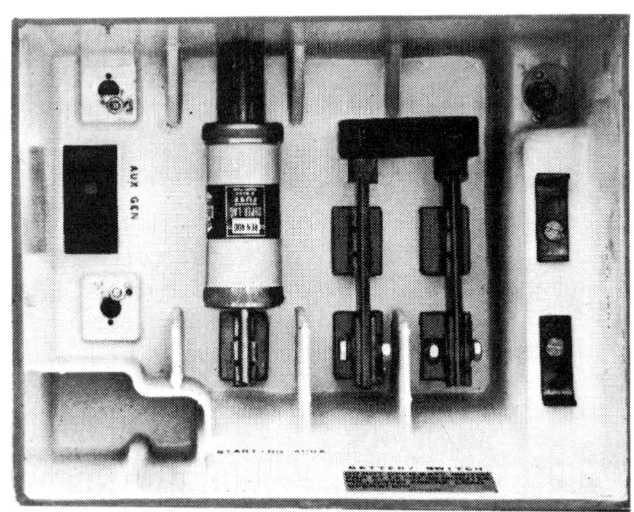

Fig. 2-23 Fuse and Switch Panel

Fig. 2-26 Governor Low Oil Trip Plunger and Engine Overspeed Trip Reset Lever

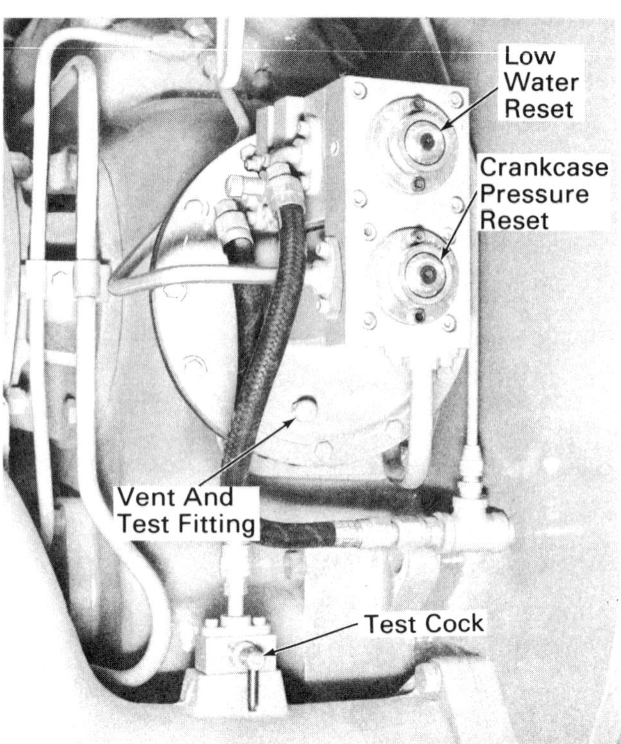

Fig. 2-27 Low Water and Crankcase Pressure Detector

Fig. 3-1 Fuel Oil Sight Glasses

OPERATOR'S MANUAL

NOTICE

The purpose of this manual is to act as a guide in the operation of the GF6C locomotive and its equipment. The information describes the equipment and operating procedures as of the time of release of the equipment.

> WARNING:
> Personnel engaged in the operation and maintenance of the locomotive must observe all the Company's directives and instructions applicable to the operation and maintenance of the high voltage equipment. Otherwise injuries might be sustained.

INTRODUCTION

To obtain the most benefit from this manual, it is recommended that the sections be read in the sequence in which they appear.

This manual has been prepared to serve as a guide to the personnel engaged in the operation of the General Motors Model GF6C Electric Locomotive. The description and operating instructions are divided into five sections as follows:

1. General Description – Provides general description of main equipment, components and systems.
2. Machine Room Operating Controls – This section briefly describes the functions of controls, indicators and devices located in the machine room.
3. Cab Operating Controls and Indicators – This section briefly describes the functions of the controls, indicators and devices located in the cab.
4. Normal Operating Procedures – This section provides information related to normal operation of the locomotive.
5. Unusual Operating Conditions – This section covers operational problems that may occur on the road and suggests actions that could be taken by the operator in response to difficulties.

Figures are identified by section and sequence. For example: Fig. 2-3 is the third figure in Section 2. Some of the figures are located the end of this manual.

TABLE OF CONTENTS

GLOSSARY

A	Ampere
AC	Alternating Current
APL	Auxiliary Power Locomotive
ATC	Automatic Train Control
BC	Brake Cylinder
B.P.C.O.C.	Brake Pipe Cut-Out Cock
DC	Direct Current
DS	Traction Motor Disconnect Switch
E.P.	Electro-Pneumatic
Hz	Hertz
km/h	Kilometers Per Hour
kV	Kilovolt
kVA	Kilovolt Amperes
LA	Lightning Arrestor
mph	Miles Per Hour
MU	Multiple Unit
OCP	Open Circuit Protection
PCS	Pneumatic Control Switch
p.f.	Power Factor
psi	Pounds Per Square Inch
S7 Cabinet	S7 Electrical Cabinet
Y1 Cabinet	Y1 Thyristor Convertor Cabinet
Y2 Cabinet	Y2 Electronics Cabinet

Note

Throughout this manual the International Standard units or S.I. units are used; for convenience and to avoid confusion, the Imperial Units are also given in brackets.

Furthermore, measurements relating to the locomotive are given in millimeters (mm), and those relating to distances are given in meters (m) or kilometers (km).

GENERAL DESCRIPTION

Model designation GF6C

Locomotive type . . . . C-C

Main Transformer – mineral oil cooled – Primary voltage 50kV/60Hz

Thyristor convertor mineral oil cooled . . . . Thyristor control system

Continuous rating . . . . 3,800 kW

Maximum diesel equivalent HP 6,000

APL power rating . . . . 270 kW

Traction motors 6
- Model E88
- Type . . . . Separately excited TM field, roller bearing suspension, grease lubricated journal bearings
- Maximum continuous tractive effort . . . . 400 kN at 34.5 km/h (90000 lb at 21.5 mph)
- Current rating maximum continuous . . . . 1042 A/motor

Driving wheels . . . . 6 pairs
- Diameter . . . . 1066.8 mm (42″)
- Gear Ratio 83:18
- Maximum speed – full kW . . . . 90 km/h (56 mph)
- – dead in tow . . . . 105 km/h (65 mph)

Pantographs 2
Make . . . . Brecknell-Willis (Ringsdorf)
Type . . . . Single Knuckle, air up, spring down

Air compressor (main) 1
Type . . . . Rotary screw
Lube oil capacity . . . . 102.3 L (22.5 I Gal)
Displacement half speed 900 RPM . . . . 1.5 m^3/min (56.2 cu.ft./min)
full speed 1800 RPM . . . . 3.8 m^3/min (135 cu.ft./min)

Air compressor (auxiliary) 1
Type . . . . Piston
Lube oil capacity . . . . 0.38 L (0.084 I Gal)
Displacement . . . . 0.079 m^3/min (2.8 cu.ft./min)

Air brakes . . . . 26 LUM

Storage batteries . . . . Lead/Calcium
Number of cells 32
Voltage . . . . 64 V
Rating (8 hours) . . . . 280 AH

Sand, total capacity . . . . 1.35 m^3 (40 cu.ft.)

WEIGHTS

Note : The weights listed below are approximate only and are intended as a guide in determining the handling procedures to be used during maintenance or rebuilding.

	kg	lbs
Total weights		
Maximum	180,859	398,720
Maximum/axle	30,445	67,118
On drivers	100%	100%
Transformer	16,136	35,500
Thyristor convertor cabinet	3,500	7,700
S1, S2 Cabinet assembly	555	1,220
S7 Cabinet assembly	1,773	3,900
Traction motor blower assembly	583	1,284
Truck assembly	27,360	60,200
Traction motors	2,722	6,000
Axle	610	1,340
Wheel	510	1,120
Axle gear 83-tooth	235	518
Rotary air compressor module	782	1,720
Aircompressor motor	242	532
Dynamic brake fan assembly	454	1,000
Dynamic brake resistor grid	73	160
Cab heater	32	71
Storage battery	130	289

MAJOR DIMENSIONS

	mm	ft.-in.
Distance between pulling face of coupling to CL of bolster	3,861	12′- 8″
Distance between bolster CL	13,259	43′- 6″
Distance pulling face front of coupling to rear of coupling	20,980	68′-10″
Width over cab sheeting	3,048	10′- 0″
Height, top of rail to top of locked down pantographs	5,029	16′- 6″
Width over handrails	3,245	10′-7¾″
Height, top of rail to top of fully extended pantographs	7,188	23′-7″

SECTION 1
GENERAL DESCRIPTION

INTRODUCTION

The General Motors of Canada Model GF6C Electric Locomotive is illustrated in Fig. 1-1. The locomotive is designed to operate as a single unit or in a multiple unit consist of two or more locomotives, up to a maximum of 6 units.

Power for operating the locomotive is obtained from the overhead catenary contact wire at an operating voltage of 50 kV, 60 Hz through a pantograph and transmitted to the primary winding of the main transformer through the main circuit breaker.

The main transformer is of a fixed ratio with multiple secondary windings. Output from the secondary windings is rectified and regulated by the thyristor convertor and smoothed by the reactors to provide direct current to the traction motor armatures. The other end of the primary winding of the main transformer is connected to a grounding transformer. Special grounding brushes on the axles are connected to the secondary of this transformer, thus providing the grounding circuit.

The traction motor field windings are connected across another secondary winding of the main transformer through thyristor controlled circuits.

The transformer windings and smoothing reactors, located in the main transformer tank, are cooled by circulating mineral oil. The thyristor controlled convertors for armature current, field current and auxiliary power are located in the Y1 thyristor convertor cabinet. The thyristors and rectifying diodes of the convertor units are mounted on heat sinks in unitized racks and are oil cooled through the heat sinks.

The cooling oil system for the transformer and reactors is separated from the thyristor convertor cooling oil system. The oil from both systems is pumped through separate heat exchangers centrally located in the Y1 thyristor cabinet. Air taken in through vents in the roof is forced down through the heat exchangers and discharged beneath the locomotive through deflectors.

The six traction motors are forced air cooled, separately excited, direct current motors. Two motor blowers provide the cooling air, taken from the roof sides and forced through inertial filters, then ducted through to the traction motors. Part of the cooling air is forced through paper filters into the electrical cabinets.

Power for operating auxiliary equipment on the locomotive is provided by an Auxiliary Power Convertor (APL). Auxiliary equipment includes the transformer cooling oil pump motor, thyristor cooling oil pump motor, traction blower motors, convertor radiator blower motors, rotary air compressor drive motor, thyristor internal blower motors and the air compressor heat exchanger blower motor.

Power is taken from one auxiliary winding of the main transformer and applied to the APL convertor in the Y1 thyristor cabinet. The auxiliary power convertor rectifies this power and inverts it to 3-phase, 480 V, 60 Hz. This power is applied to the auxiliary equipment as required.

Power for the battery charger, cab heaters and equipment room heaters is taken from the same auxiliary winding in the main transformer. The voltage is stepped down by an auxiliary transformer, then applied to the battery charger (120 V) and cab heaters (240 V).

LOCOMOTIVE GENERAL ARRANGEMENT

The location of components is shown in the locomotive general arrangement illustration, Fig. 1-1.

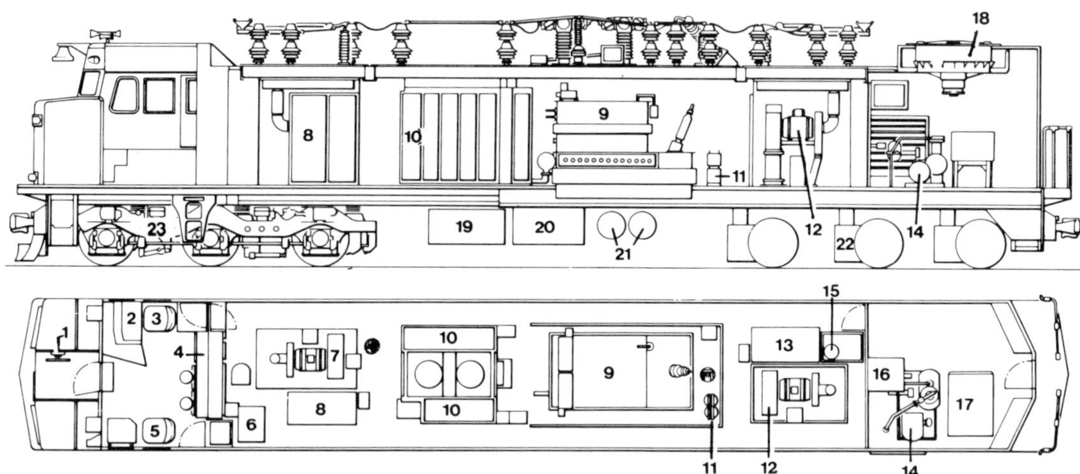

1. Handbrake
2. Operator's Console
3. Operator's Seat
4. S7 Panel
5. Assistant's Seat
6. Radio Equipment
7. Traction Motor Blower
8. Contactor Cabinet
9. Main Transformer
10. Thyristor Convertor Cabinet
11. Auxiliary Transformer
12. Traction Motor Blower
13. Contactor Cabinet
14. Air Compressor
15. Auxiliary Compressor
16. Dynamic Brake Contactors
17. Compressor Cooling
18. Dynamic Brake Fan
19. Capacitor Box
20. Filter Reactor Box
21. Main Reservoirs
22. Traction Motor
23. Truck

Fig. 1-1 General Arrangement

SYSTEMS

A brief description of the major systems of the locomotive is provided in the following paragraphs.

Pantographs

A single knuckled pantograph is located at each end of the locomotive. Each pantograph is directly connected to the main circuit breaker. BOTH PANTOGRAPHS ARE ENERGIZED WHEN EITHER ONE IS RAISED. They are air operated, air raised, spring lowered. Both electric and pneumatic selector switches have to be set correctly to raise the chosen pantograph. Only one pantograph can be raised at any given time.

After having selected the correct pantograph and set the pneumatic and electrical switches on all the locomotives of the consist, actuating the "pan" switch (5) on Fig. 2-20, on the control desk in the control cab, will raise all the pantographs. All pantographs can be lowered by placing the "pan" switch in the down position.

A protection system maintains a minimum voltage on the batteries to ensure enough energy for raising a pantograph.

Grounding Switch

A hand operated grounding switch contacts both sides of the main breaker grounding the high voltage side of the power system for safe maintenance or repairs. A safety interlock will prevent the pantographs from being raised when the grounding switch is in the GROUNDED position.

Thyristor Convertor

The thyristor convertor cabinet consists of equipment to control the armature and field currents of the six traction motors and supplies three-phase power for all auxiliary motors. The convertor consists of three units. One centrally placed air/oil and air/air heat exchanger cooling unit and two convertor cabinets situated on each side of the heat exchanger.

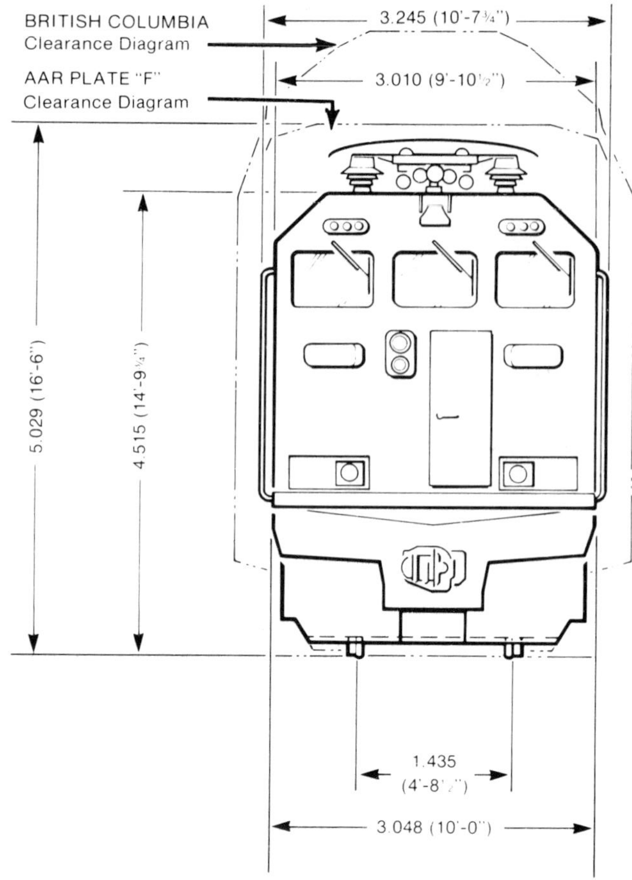

Fig. 1-2 General Arrangement (front)

Power Factor Correction

Power factor correction is achieved by means of thyristor switched capacitor reactor banks connected in parallel across the four rectifier bridges used to control traction motor armature current. The capacitor bank and reactor bank are located under the locomotive underframe.

Auxiliary Power Convertor

Power from the auxiliary power system is 3-phase, 480 V, 60 Hz, 232 kW, or 3-phase, 240 V, 30 Hz, 116 kW depending upon operating conditions. The Auxiliary Power Convertor provides 3-phase power for the auxiliary equipment consisting of the transformer and thyristor convertor cooling oil pump motors, traction motor blower motors, convertor radiator blower motors, an air compressor motor, an air compressor heat exchanger blower motor and thyristor convertor internal blower motors.

Air Compressor

Situated towards the rear end of the locomotive, a rotary screw compressor supplies all the compressed air required during normal operation. Its maximum rated output is 3.8 m^3/min (135 cu.ft./min). This compressor is cooled by circulating the oil through an air cooled radiator. This cooling air can be exhausted outside for summer operation and inside for winter.

Auxiliary Air Compressor

The auxiliary air compressor supplies the necessary air to raise the pantograph and operate the main breaker for startup operations when the main air pressure is insufficient.

Air Brake System

The locomotive is equipped with 26LUM air brakes, including a model 30-CDW brake valve. This air brake equipment is situated under the cab floor. The controls are on the operator's console directly in front of the operator. The palm countoured lever handles operate in the vertical plane and are detented to provide a positive feel and location of the lever.

Dynamic Brake

The dynamic brake lever is the same handle as the throttle control lever. Pushing this lever forward beyond the Dynamic Brake set-up notch will progressively apply the dynamic brake. The grids are mounted radially with one cooling fan in a hatch at the rear end of the locomotive. The grid current is regulated at 760 A.

Pacesetter Control
This automatic control will maintain an adjustable constant speed for loading operations. This speed is maintained irrespective of load conditions.

Battery Charger
The locomotive lighting and low voltage control system operates on 74 VDC. A separate secondary winding on the auxiliary transformer supplies AC current to a thyristor controlled battery charger. This charger has a flat output curve limiting the output to exactly 100 A at 74 VDC.

Battery Protection
The battery charger is also equipped with a protection system that prevents discharge of the batteries below 55 VDC. This is sufficient to operate the auxiliary compressor for raising a pantograph. The battery charger is also equipped with an override switch.

Fire Protection
Fire detectors are fitted in all sections of the locomotive and in all the electrical cabinets. "HALON" gas bottles are located in the machine room and the gas is piped to various nozzles. In case of a fire, there is a 15 second time delay to allow the machine room area to be evacuated after the warning siren is activated, and before the gas is actually released; lights will also flash a warning at the machine room entrance doors in the control cab and compressor compartment.

Wheel Creep Control System
The wheel creep control system used on this locomotive allows each wheel to achieve maximum tractive effort by allowing the wheels to creep (rotate slightly faster than ground speed) at a controlled rate. Under severe rail conditions, sanding may be applied automatically to maximize wheel to rail adhesion. At speeds greater than 8 km/h (5 mph) manual sanding is disabled, as long as the creep control system is operating and the locomotive is being operated in the power mode.

If and when a creep control system failure occurs, the locomotive control system will detect the failure and transfer wheel slip control to the backup system, which is a conventional rate of change and level detection wheel slip correction system, utilizing power reductions and automatic sanding applications.

When the locomotive is operating with the "backup" system, the wheel slip light and the buzzer located on the control console operate as visible/audible indications of wheel slip. These two indicators are not functional when operating in the creep control system mode.

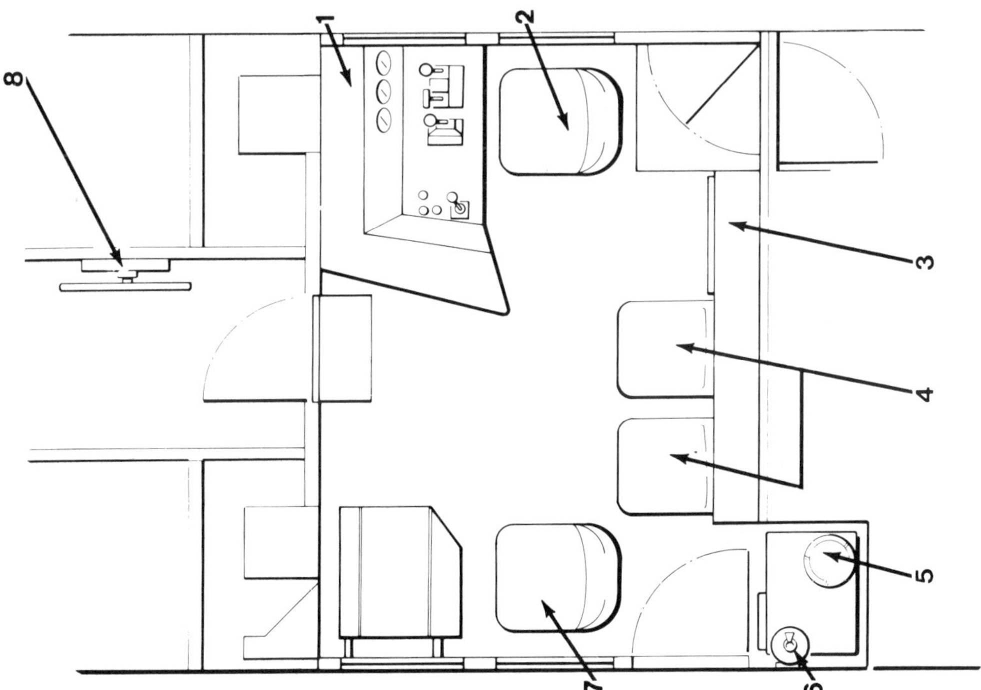

1. Control Console
2. Operator
3. S7 Panel
4. Seats
5. Water Cooler
6. Fire Extinguisher
7. Assistant
8. Hand Brake

Fig. 1-3 Cab Arrangement

SECTION 2
CONTROLS AND INDICATORS IN MACHINE ROOM AND CONTROL ROOM

This chapter lists, locates and provides a brief description of the functions of all operating controls in the machine room and the control cab of the GF6C Electric Locomotive.

Machine Room
- Pantograph Panel, Fig. 2-1
- Others.

Control Cab
- The S7 electrical panel with the following sections:
  - Breaker panel 74 VDC, Fig. 2-4
  - Breaker panel 120-240 VAC, Fig. 2-5
  - Locomotive control panel, Fig. 2-6
  - Breaker panel 480 VAC, Fig. 2-8
  - Battery Switch, Fuse and Test panel, Fig. 2-9
- Main Control Console subdivided as follows:
  - Front vertical panel, Fig. 2-11
  - Front horizontal desk, Fig. 2-12
  - Side vertical panel, Fig. 2-16
  - Lower part panels, Fig. 2-18
- Overhead Controls
  - Operator's Side, Fig. 2-19
  - Passenger's Side.
- Miscellaneous

MACHINE ROOM & CAB CONTROLS

MACHINE ROOM

Pantograph Panel
Situated over the auxiliary compressor, this panel contains the main controls to activate the locomotive auxiliary compressor and to set up the circuits to raise the pantograph.

Electric Selector Switch
This three position switch should be in the OFF position when the pantograph is down. "Hood end" or "cab end" pantograph to be raised can be selected by placing the switch in the appropriate position.

Pneumatic Selector Switch
The two position pantograph selector valve lever must also be placed in the same position as the electric selector.

Auxiliary Compressor Start Button
If the pressure in the compressed air tank is insufficient to raise the pantograph, starting the auxiliary compressor with the reserve power of the batteries will supply the necessary air pressure to raise and maintain the pantograph until the main air compressor takes over. An automatic pressure sensor will cut out the auxiliary compressor at that point.

MISCELLANEOUS
Thyristor Convertor Oil Level

Fig. 2-2 Oil Level Indicators

An oil level dial type indicator is located on the right hand side of the convertor at eye level. Oil level should be maintained in the green area located between "min" and "max".

Transformer Oil Level
An oil level dial type indicator is located on the right hand side of the main transformer at eye level. Oil level should be maintained betwen "min" and "max" readings at 20 deg. C (68 deg. F) temperature.

Grounding Switch (Fig. 3-1)
A grounding switch is provided on the left side of the machine room. When this handle is in the grounded position, both sides of the Main Circuit Breaker will be grounded. Refer to page 3-2 for safe operation.

WARNING: Never attempt to open any access to the high voltage enclosures without having opened the main breaker, lowered the pantograph and placed the grounding switch in the "grounded" position. Engagement of the grounding switch on the roof must be visually confirmed.

Heater Thermostats
Two thermostats at each end of the machine room control the heaters when the heater control switch on the locomotive panel is ON.

Humidity Indicators
A humidity indicator is situated on the auxiliary compressor showing the level of moisture content of the compressed air. The indicator colour will change from blue to white as the moisture level increases. When correct humidity operating conditions prevail, the indicator should be BLUE. If the indicator is completely WHITE the system could be water saturated and must be checked.

Two more indicators are located on the main air compressor dryers, situated below the underframe on the left side of the locomotive. The indications are the same as above.

Light Switches
One machine room light switch is situated at each end of the room, just inside the access doors.

Each electrical cabinet in the machine room has a light switch situated just inside the panel door. The main cabinet has an external switch located on the right hand side of the cabinet.

CAB CONTROLS – S7 PANEL (Fig. 2-3)
This main electrical control cubicle is situated directly behind the operator's seat. On the cab side it contains the main circuit breakers and locomotive control panels. Illustration 2-3 indicates the position of the secondary electrical panels.

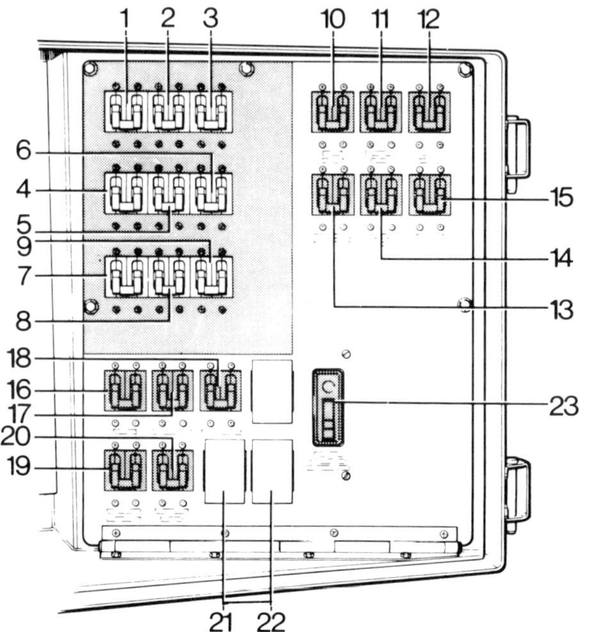

1. ★Battery Protection Circuit
2. ★Control
3. ★Local Control
4. ★Motor Disconnect Switch Protection 1, 2, 3
5. ★Motor Disconnect Switch Protection 4, 5, 6
6. ★Electric Supply Voltage
7. ★Auxiliary Air Compressor
8. ★Transfer Control I
9. ★Transfer Control II
10. Voice Radio
11. Locotrol Radio
12. LIC Radio
13. Snow Plow Headlights
14. Ditch Light
15. Layover
16. Lights
17. Headlights
18. Pacesetter
19. Automatic Blow Down Timer
20. Air Dryer Heater
21. Battery Charger

Fig. 2-4 74 VDC Panel

BREAKER PANEL 74 VDC (Fig. 2-4)
This panel comprises the main control breakers. These breakers can be controlled manually as switches and will automatically trip in case of an overload. The section painted black on the panel and marked by an asterisk in the following description must be in the ON position to operate the locomotive. Other breakers can be placed as required.

(1) **★Battery Protection Circuit 5 A**
Provides a minimum safe level of discharge of the batteries.

(2) **★Control 40 A**
This breaker protects the 13T control line. All controls, including the pantograph, are dead when this circuit is open.

(3) **★Local Control 30 A**
This circuit breaker protects the locomotive positive (PA) and negative (NA) local control circuits operating heavy duty switch gear and various control devices.

(4) **★Motor Disconnect Switch Protection 1, 2, 3 3 A**
Protects the motorized disconnect switching circuits for traction motors no. 1, 2, and 3 (truck 1).

(5) **★Motor Disconnect Switch Protection 4, 5, 6 3 A**
Same as above but for truck 2.

(6) **★Electronic Supply Voltage 15 A**
Electronic control circuits are assembled on plug-in circuit modules to facilitate maintenance. 74 VDC power is supplied to the electronics cabinet Y2 through this circuit breaker which protects the local control circuit to the boards.

(7) **★Auxiliary Air Compressor 20 A**
Protects the auxiliary compressor circuits. Must be ON if the pantograph is to be raised when the main reservoir air pressure is low.

(8) **★Transfer Control I 3 A**
Provides protection for the motoring to dynamic braking transfer switch circuit for the no. 1 truck.

(9) **★Transfer Control II 3 A**
Same as above for the no. 2 truck.

(10) **Voice Radio 6 A**
Protects the circuits of the communication radio situated on the control console.

(11) **Locotrol Radio 6 A**
Protects the circuit supplying the radio link required for controlling a locomotive further down in the consist during locotrol operation.

(12) **LIC Radio 6 A**
Protects circuits to the radio system for automatic location identification and control system.

(13) **Snow Plow Headlights 30 A**
Protects the circuits of the snow plow headlights.

(14) **Ditch Light 20 A**
Protects the circuits of the ditch lights.

(15) **Layover 15 A**
Protects the control circuits associated with the layover control system.

(16) **Lights 30 A**
This circuit breaker protects the circuit to the front and rear cab lights, desk lights; gauge lights, number lights, class light, and boarding lights for cabs. This circuit breaker also protects the circuit for the 74 VDC receptacle in the cabs and equipment room and to the equipment room lights.

(17) **Headlights 35 A**
This circuit breaker protects the circuits of the front and rear headlights of the locomotive.

(18) **Pacesetter 15 A**
This breaker ensures the protection of the pacesetter equipment.

(19) **Automatic Blow Down Timer 15 A**
This circuit breaker provides protection for the compressed air system automatic drain valve timer circuit.

(20) **Air Dryer Heater 15 A**
This circuit breaker protects the circuits for the air dryer heater on the compressed air system.

(21) **Battery Charger 150 A**
This circuit breaker provides the protection for the battery charging circuits when the batteries are being charged from an external receptacle.

BREAKER PANEL 120/240 VAC (Fig. 2-5)

(1) **★Battery Charger 100 A**
This breaker protects the input to the battery charger system from the auxiliary transformer. This breaker should be in the ON position at all times except during maintenance or repairs.

(2) **★Catenary Voltage 5A**
Protects the control voltage detection circuit, absence of which reduces the locomotive power to zero. This breaker must be on for the locomotive to be operational.

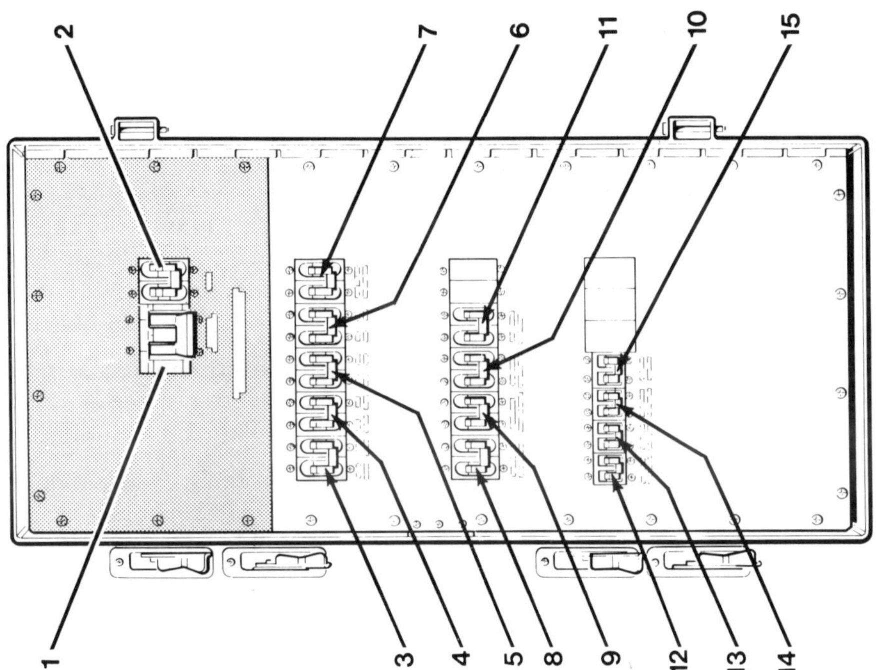

1. ★Battery Charger
2. ★Catenary Voltage
3. Cab Heater Engineer Side
4. Cab Heater Assistant Side
5. Side Heater Engineer Side
6. Side Heater Assistant Side
7. Battery Pad Heater
8. Machine Room Heater
9. Machine Room Heater
10. Compressor Oil Heater
11. Machine Room AC Receptacle
12. Hot Plate
13. Window Heater
14. Mirror Heater
15. Water Cooler

Fig. 2-5 120/240 VAC Panel

(3) **Cab Heater Engineer Side 15 A**
This breaker protects the heater on the engineer's side of the cab. The capacity of this heater is 2 kW.

(4) **Cab Heater Assistant Side 30 A**
Same as above but for assistant side. This heater is 5 kW.

(5) **Side Heater Engineer Side 10 A**
This heater is situated at the right side of the operator and has a capacity of 1.5 kW.

(6) **Side Heater Assistant Side 10 A**
Same as above but to the left side of the cab.

(7) **Battery Pad Heater 10 A**
Protects the heating circuit to the battery pad, during normal operation or during layover from outside supply. Capacity is 953 W.

(8) **Machine Room Heater 70 A.**
The two heaters are in operation when switched on from cab control ref. 8, p. 2-13 during layover periods. These heaters are controlled by a thermostat in the machine room and have a capacity of 5 kW each.

(9) **Machine Room Heater 70 A**
Same as above but on the other end of the machine room.

(10) **Compressor Oil Heater 10 A**
Protects circuit to the compressor oil heaters for maintaining adequate oil temperature during layover periods. Circuit may be supplied from an outside or inside source.

(11) **Machine Room AC Receptacle 15 A**
Protects the 120 VAC power supply available through the receptacles in the machine room.

(12) **Hot Plate**

(13) **Window Heater**

(14) **Mirror Heater**

(15) **Water Cooler**

LOCOMOTIVE CONTROL PANEL (Fig. 2-6)
This panel is located on the S7 cabinet at the top left corner. It contains the following switches:

(1) **Dynamic Brake Cut-out Switch**
The switch situated at the top left corner will CUT OUT the dynamic brake on the unit. This switch can be used to limit the number of units in the consist with dynamic brake capability. This switch can also be used for trouble-shooting purposes or for cutting out a defective unit.

(2) **Locked Wheel Cut-out**
For the locked wheel detection system to be operational, this switch must be in the OFF position. This system will be in operation at all times. In the ON position the system is nullified.

(3) **Traction Motor Cut-out Switches**
The six (6) motor cut-out switches must be in the OFF position for normal operation. If a fault develops in any of the traction motors, the corresponding switch can be placed in the ON position and the motor cut out. Dynamic braking is not available on the unit when a traction motor is cut out.

(4) **Ground Relay Lock-out Reset (76)**
This ground fault relay will automatically reset three (3) times before it locks out. To reset, operate this switch will allow the ground relay to operate for one more cycle (3 times).

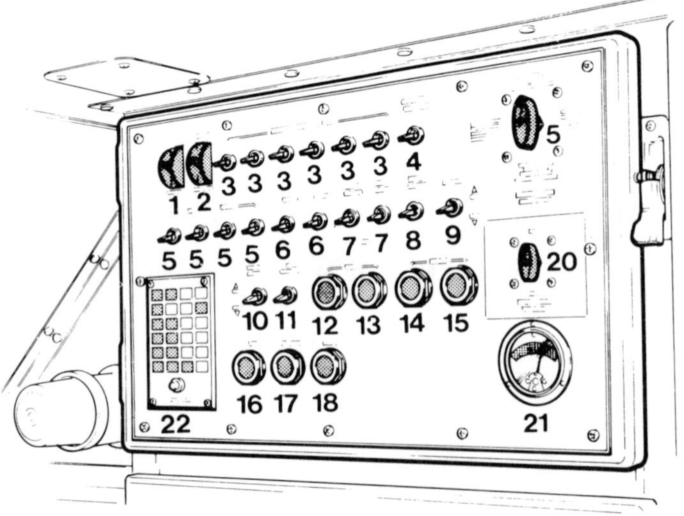

1. Dynamic Brake Cut-out Switch
2. Locked Wheel Cut-out
3. Traction Motor Cut-out Switch
4. Ground Relay Lock-out Reset
5. Power Factor On/Off Switches
6. Light Switches
7. Light Switches
8. Machine Room Heater Switch
9. Layover
10. Battery Protection Override
11. APL 1/2 Speed Nullification
12. DS Switch 1, 2 and 3
13. DS Switch 1, 2 and 3
14. DS Switch 4, 5 and 6
15. DS Switch 4, 5 and 6
16. OCP Reset
17. (86) Relay Reset
18. Battery Protection Reset
19. Headlights
20. Isolation Switch
21. Battery Ammeter
22. Indicator Lights

Fig. 2-6 Locomotive Control Panel

(5) **Power Factor On/Off Switches (4)**
Situated on the left side of the panel, these four switches connect the power factor correction system across the input to each thyristor bridge. During normal operation these switches must be ON. Switch off a circuit only if a fault occurs in it.

(6) **Lights**
(7) The following four switches are to be used as required:
- the number lights front and rear,
- the platform lights,
- the step and ground lights.

(8) **Machine Room Heaters**
In normal locomotive operation this switch can remain in the OFF position. The machine room heaters are controlled by the two thermostats in the machine room and the layover heater protection circuit breaker. During layover periods, when the pantograph is down and the 240 VAC external supply is connected, this switch will control the heaters through the thermostats.

(9) **Layover**
As above, when the pantograph is down and the 240 VAC or 480 VAC external supply is connected, this switch should be placed in the ON position to energize the layover systems and lights.

(10) **Battery Protection Override**
The battery protection system will disconnect the battery if the charger is inoperative and the voltage drops below 55 VDC. Should power be required for the auxiliary compressor, or for any other reasons, the override should be switched on. Normally this switch must remain OFF.

(11) **APL 1/2 Speed Nullification**
The function of the Auxiliary Power Convertor is to supply 3-phase power for the locomotive auxiliary motors. The circuit controlled by this switch, when ON, will nullify the 1/2 speed motor operation. With the switch in the OFF position the 1/2 speed function is available. Normal position of the switch during operation is OFF.

(12) **DS Switch 1, 2 3**
(13) These two push buttons, OPEN and CLOSED, control the main disconnect switch to the traction motors 1, 2 and 3. When in the OPEN position truck no. 1 is out of operation. In normal operation this should be in the CLOSED position.

(14) **DS Switch 4, 5, 6**

(15) Same as above but for truck no. 2 or traction motors 4, 5 and 6.

(16) **OCP Reset**
After the dynamic brake system has been tripped due to an open circuit in the grids, the Open Circuit Protection relay reset will allow the system to become operative once more.

(17) **Relay Reset (86)**
This protection relay will trip the main circuit breaker whenever a fault or overcurrent occurs on the main power and traction circuits. This reset button will reset the main circuit breaker when the fault has been cleared or isolated.

(18) **Battery Protection Reset**
After the battery protection system has been tripped, this reset will allow the system to become operative once more, provided the power supply for charging the batteries is available. This battery protection override switch must be reset in the OFF position if it has been used.

(19) **Headlights**
The locomotive is equipped with twin sealed-beams at both ends. These headlights are controlled from a switch on the overhead control panel, driver's side. But before this switch can operate, the main rotary switch at the top right corner of the locomotive control panel must be set. This switch should be set in the required position depending on the locomotive position in a consist at the start of operation.

- ON LEAD UNIT
  If only a single locomotive unit is being used, place the switch in the SINGLE UNIT position, twelve o'clock.

In multiple unit operation, if trailing units are coupled to the no. 2 end, (long hood) of the lead unit, place the switch in the CONTROLLING – COUPLED AT LONG HOOD END position three o'clock.

In multiple unit operation, if trailing units are coupled to the no. 1 end (short hood) of the lead unit, place the switch in the CONTROLLING – COUPLED AT SHORT HOOD END position, six o'clock.

– ON INTERMEDIATE UNITS
On units operating between other units in a multiple unit consist, place the switch in the INTERMEDIATE UNIT position, twelve o'clock.

– ON TRAILING UNITS
The last unit in a multiple unit consist should have the headlight control switch placed in the CONTROLLED – COUPLED AT EITHER END position, nine o'clock.

(20) **Isolation Switch**
This is a three (3) position switch controlling the main status of the locomotive. These positions are STOP, STAND-BY and RUN.

– STOP
During normal layover, major break down, or when being hauled dead, the isolation switch must remain in this position. The pantograph will be down and cannot be raised.

– STAND-BY
When in this position, the pantograph can be raised and all auxiliary systems energized, but the main traction motor and systems will not operate.

– RUN
When the locomotive is ready to run, place the switch in this position. It will now be ready to move and the locomotive can now be controlled from the console.

(21) **Battery Ammeter**
This indicator measures the amount of current flowing to or from the storage batteries. Since in normal operation the batteries are well charged, the ammeter should read zero or slightly in the green area. The ammeter in the red indicates that the batteries are discharging or the cells are freezing, either of which could lead to possible difficulties at start-up.

(22) **Indicator Lights** (Fig. 2-7)
A panel of 20 indicator lights plus 4 spares at the right bottom corner of the panel give visual indication of major faults occurring in the system. For corrective measures, refer to the "UNUSUAL OPERATING CONDITIONS" section of this manual. Their functions are as follows:

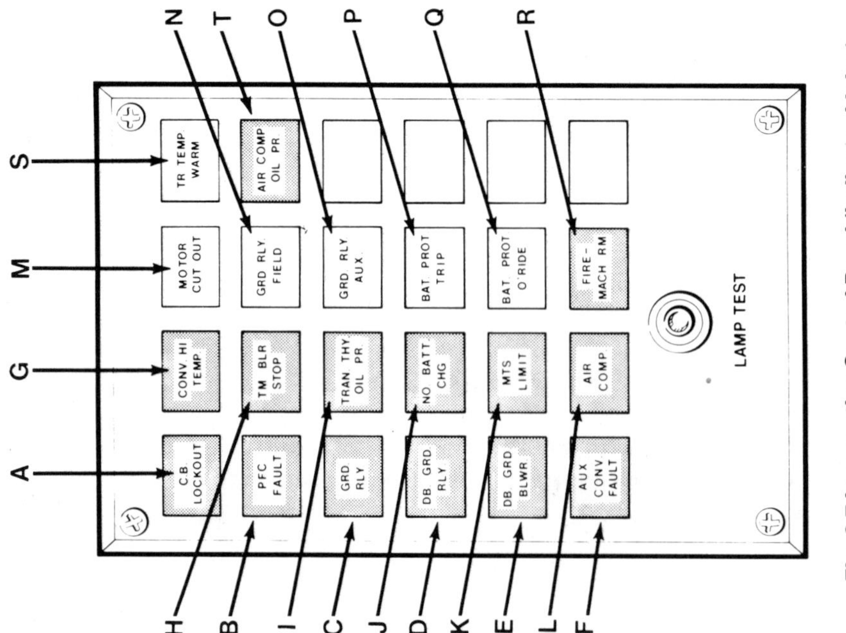

Fig. 2-7 Locomotive Control Panel (Indicator Lights)

(22A) CB LOCK-OUT (Circuit Breaker) – RED
Current in any of the main transformer secondary windings is excessive.
Transformer oil pump circuit breaker is tripped.
Fire extinguisher system is activated.

(22B) PFC FAULT (Power Factor Correction) – RED
Indicates a fault is detected in any of the four power factor correction circuits.

(22C) GRD RLY (Ground Relay) – RED
An overcurrent and/or ground fault is detected in any of the traction motor armature circuits.

(22D) DB GRD RLY (Dynamic Brake Ground Relay) – RED
Indicates a ground fault in the dynamic brake grid circuits.

(22E) DB GRD BLWR (Dynamic Brake Grid Blower) – RED
Open circuit or short circuit in the grid blower circuit.

(22F) AUX CONV FAULT (Auxiliary Convertor) – RED
Phase fault in the auxiliary convertor or the auxiliary convertor is disabled.

(22G) CONV HI TEMP (Convertor High Temperature) – RED
High air temperature in the thyristor convertor.
High oil temperature in the thyristor convertor.

(22H) TM. BLR. STOP (Traction Motor Blower) – RED
Either traction motor blower motor has stopped.

(22I) TRANS. THY. OIL. PR. (Transformer/Thyristor Oil Pressure) – RED
Low oil pressure is detected in the transformer.
Low oil pressure is detected in the thyristor convertor.

(22J) NO BATT. CHG. (No Battery Charge) – RED
Battery charger is not operating.

(22K) MTS. LIMIT (Traction Motors) – RED
Traction motor thermal capacity is reached.

(22L) AIR COMP. (Air Compressor) – RED
The air compressor circuit breaker is tripped.
Hot oil temperature in air compressor.
Cold oil temperature in air compressor.

(22M) MOTOR CUT-OUT (Traction Motor) – WHITE
One or more traction motors are cut out.

(22N) GRD. RLY. FIELD (Ground Relay) – WHITE
A ground fault is detected in the traction motor field windings.

(22O) GRD. RLY. AUX. (Ground Relay Auxiliary) – WHITE
A ground fault is detected in the auxiliary power convertor circuit.

(22P) BAT. PROT. TRIP. (Battery Protection) – WHITE
The battery voltage has decreased below 55 volts and the battery charger is not operating.

(22Q) BAT. PROT. O'RIDE (Battery Protection Override) – WHITE
Indicates the BATT. PROTECTION OVERRIDE switch is in the ON position.

(22R) FIRE MACH. RM. (Machine Room) – RED
Indicates a fire is detected in the machine room.

(22S) TR. TEMP. WARM (Transformer Temperature) – WHITE
Indicates a warm oil temperature in the main transformer.

(22T) AIR COMP. HOT OIL (Air Compressor) – RED
Indicates an overtemperature of the compressor oil.

NOTE: See page 4-5 for corrective information.

BREAKER PANEL 480 VAC (Fig. 2-8)
This panel is situated in the center of the S7 cabinet and all its breakers must be ON for the locomotive to be operational. It contains the following breakers:

(1) **★ Thyristor Convertor Radiator Blowers 18 A**
These two breakers protect the radiator blower motors forcing air through the central cooling unit located within the thyristor convertor for cooling the convertor and transformer oil as well as the air within the thyristor convertor cabinet.

(2) **★ Traction Motor Blowers 225 A**
These two breakers protect the two traction motor blower motors situated at each end of the machine room. The blowers supply cooling air to each truck. These blowers also provide air to pressurize the main electrical cabinets.

(3) **★ Y1 Internal Blowers 2 A**
These two breakers protect the two blower motors used for circulating air within the thyristor convertor cabinet.

(4) **★ Thyristor Convertor Oil Pump 7 A**
This breaker protects the oil pump motor circulating the convertor oil through the central cooling unit.

(5) **★ Air Compressor Motor 100 A**
This breaker protects the main air compressor motor supplying air to the compressed air system.

(6) **★ Transformer Oil Pump 12 A**
This breaker protects the oil pump motor circulating the transformer oil through the central cooling unit.

(7) **★ Air Compressor Oil Cooler 6.2 A**
This breaker protects the oil cooler fan motor for the main air compressor. This oil is circulated by the compressor itself.

SWITCH AND FUSE PANEL (Fig. 2-9)
This panel is located at the left of the S7 cabinet behind the operator, below the locomotive control panel. It contains the following equipment:

Fuse Test Equipment
To facilitate the testing of fuses, a pair of fuse test blocks, a test light and a test light toggle switch are installed on the fuse panel. To test fuses proceed as follows:

- Place the switch ref. 7 in the ON position to test the light bulb ref. 8. It should light up; if not, change the bulb.
- Switch back the toggle to the OFF position and place the metal ends of the fuse to be tested firmly against the two metal test blocks ref. 9.
- If the fuse is good, the light will come on.

It is always advisable to test fuses before installing them.

CAUTION: REMEMBER TO ALWAYS ISOLATE THE CIRCUIT BEFORE REPLACING A FUSE.

(1) **Main Battery Fuse 150 A**
This is the first fuse from the left. It protects the main battery circuit.

(2) **Auxiliary Transformer Fuse 100 A High Voltage**
This is the middle fuse. It protects the auxiliary transformer circuit.

(3) **Auxiliary Power of the Locomotive 450 A High Voltage**
This is the large fuse on the right. It controls the supply from the secondary winding of the main transformer to the APL system.

(4) **Ground Relay Cut-out Switch**
The purpose of the ground relay cut-out switch is to eliminate the ground protective relay from the locomotive circuits during certain shop maintenance procedures. IT MUST ALWAYS BE IN

THE "RUN" POSITION during normal operation. Also, the ground fault must be corrected before operation is resumed with this locomotive.

(5) **Battery Switch**
The large rotary switch is the main battery switch. It connects the batteries to the locomotive low voltage system and should be kept in the "ON" position at all times during operation.

(6) **74 Volt Receptacle**
A receptacle mounted on the switch and fuse panel makes 74 VDC available for maintenance or test purposes. The power is present at the receptacle when the battery rotating switch is ON and the 30 A LIGHTS circuit breaker is closed.

LOCOMOTIVE CONTROL CONSOLE

(Figure 2-10)

The GF6C Locomotive control console contains all the necessary controls and indications for the operation of the locomotive. All these controls and instruments are described below with a brief indication of their function and operation; for more details on their operation refer to the next chapter starting at page 3-1.

To facilitate orientation, this console and the immediate area under the control of the operator have been divided into seven different sections, four on the console itself and three in the immediate surroundings, these are:

Main Console

- Front Vertical Panel
- Front Horizontal Panel
- Side Vertical Panel
- Lower Area

Overhead Panels

- Operator's Side
- Assistant's Side

Miscellaneous

FRONT VERTICAL PANEL (Fig. 2-11)
This is the panel directly facing the operator, it comprises three main groups of controls and instruments from left to right:

Switches
Three toggle switches control the following lighting circuits:

(1) **GAUGE LIGHTS** for all the gauges of the console, the corresponding dimmer is situated on the lower panel at the left of the operator.

(2) **READING LIGHT** which will project light on the desk top for reading or writing.

(3) **DITCH LIGHTS** will project light to the front left and front right of the locomotive.

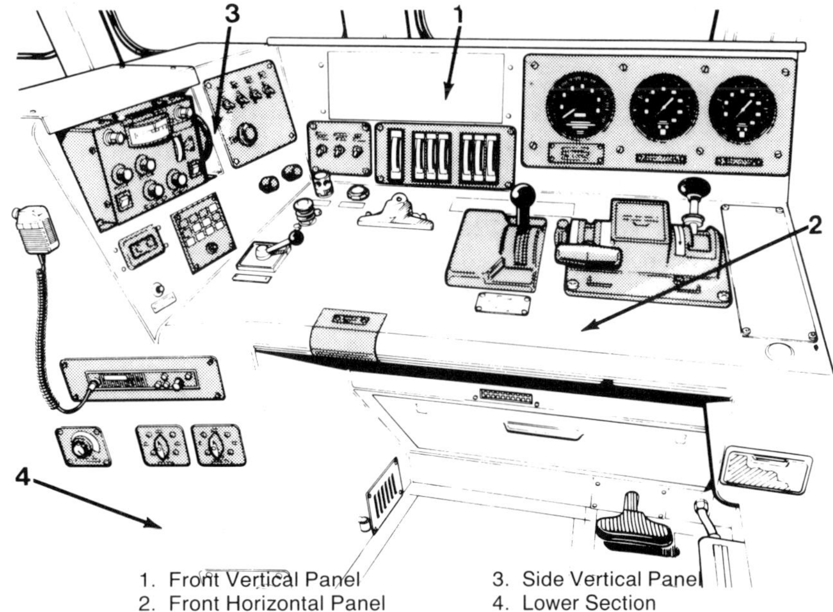

Fig. 2-10 Control Console

Ammeters
The center panel comprises seven ammeters, one independent and two groups of three. They are:

(4) DYNAMIC BRAKE AMMETER. The first ammeter on the left is graduated from 0 to 1000 A indicating the braking effort generated by the traction motors and dispersed by the dynamic brake grids. See "Dynamic Braking".

(5) TRACTION MOTOR AMMETERS. These six ammeters are grouped by truck, showing the relative tractive effort of each motor. They are graduated from 0 A to 1700 A. The traction motors are rated from 1042 A maximum continuous operation.

Should the motor temperature rise above a safe limit due to high current, an automatic control from the traction motor temperature simulator will reduce power to the continuous operating rating of 1042 A.

NOTE: If a diesel locomotive is part of the consist it is not protected by the traction motor temperature simulator and it is left to the operator to limit the time of full power application. This applies also in case of failure of the simulator. The following values can be used as guidelines and are typical only for the SD40-2 locomotives. The exact values will depend on the specifications of the diesel locomotive involved:

1050 A for continuous operation
1075 A not over one hour
1100 A not over 1/2 hour
1150 A not over 1/4 hour

Gauges
The panel on the right of the vertical front panel includes a speed indicator and two duplex compressed air gauges.

(6) SPEED INDICATOR. Is situated directly in front of the operator and is graduated from 0 to 75 mph (120 km/h). This reading is obtained from an electronic signal processed from the radar measurement of actual track speed. Operating speed of the locomotive at full power is limited to 90 km/h (56 mph). The locomotive may be safely towed dead up to 104 km/h (65 mph).

(7) AIR GAUGES. These two gauges indicate four different pressures in the air brake system. Each gauge has two needles of different colors, the code being indicated over the dials.

The first dial on the left indicates the pressure in the main reservoir (red needle) and the equalizing reservoir (white needle).

The second dial on the right indicates the pressure in the brake cylinder (red needle) and the pressure in the brake pipe (white needle).

Location, Identification and Control System (LIC)
The LIC system is a computer based block control system that reports back to central dispatch the locomotive speed, location and direction and enables direct dispatching and signalling.

Onboard electronic equipment receives and stores information from transponders located in the roadbed; then relays this information to the dispatching office. From there instructions can be transmitted back to the locomotive cab display unit, located in front of the operator on the control console.

FRONT HORIZONTAL PANEL (Ref. 2-12)
This panel is at the top of the operator's console directly in front of him. It comprises the more important driving controls, namely from left to right and top to bottom:

(9) **Alarm Silencer**
In the event of a visual and audible alarm being given by the control system, this push button will silence the audible alarm leaving "ON" the visual alarm on the light panel until the fault has been corrected and/or reset. A light on the switch itself will light up indicating the alarm has been silenced.

NOTE: In this mode a new fault or problem will not be announced by the bell.

(10) **Manual Sanding**
Once the wheel slip control has detected slippage of the wheels, it will modulate the power output of each traction motor. No manual sanding is available over 5 km/h. If sanding is required during starting, this push button will supply sand in front of each truck.

(11) **Bell Switch**
Will switch on the locomotive bell when required during movement in yards, or as per company regulations.

(12) **Reverser Control**
This handle is situated at the extreme left of the desk. It has three detent positions: when in the center position, the control is in the NEUTRAL position, the interlocks are in effect, and the unit will not be able to move under power. When the handle is moved FORWARD, the locomotive will move in that direction. When the handle is moved to REVERSE, the locomotive will move in the reverse direction.

When the reverser handle is removed, the throttle controls are electrically interlocked and cannot be operated. Note that the throttle handle can still be moved, but this will have no effect on the operation of the locomotive.

(13) **Throttle Handle and Dynamic Brake**
The throttle handle is the first on the left of the three main handles in front of the operator. It is combined with the dynamic brake into one integrated operation. The "O" position is centered at the vertical gate offset.

- THROTTLE – To operate the traction motors in power, PULL the throttle lever towards the operator. There are eight (8) different power positions, but the lever will move smoothly from "O" to full power.
- GATE – When the throttle lever is in power, to bring it back to the "O" position, push it forward until it reaches a definite stop. To pass into dynamic braking the handle must be RAISED above the detent before being able to move further forward.
- DYNAMIC BRAKE – To operate the dynamic brakes, from the traction range, or from zero traction stop, raise the lever over the stop and push it forward through the full range to obtain maximum braking effect at 760 A.

Air Brake Equipment
The air brake equipment comprises a group of controls directly at the right side of the throttle. The three valves of this block are:

(14) Brake Valve Cut-Off Pilot

(15) Automatic Brake

(16) Independent Brake

The 30-CDW automatic and independent WABCO valve system combines the three controls in a streamlined housing conserving space and simplifying the operation. The palm contoured handles, utilizing the vertical plane for operation, initiate brake control commands similar to those of conventional brake systems.

Significant operation positions are detended for positive feel and location. Both handles have an individual latching arrangement that allows handle removal for safety reasons.

Brake Cut-Off Valve (Fig. 2-13)
A latching detent provides manual operation of the cut-off valve and may be placed in three positions: "Out", "Frt" and "Pass".

- OUT, signifies that the brake system is not operational but can be tested for servicing or for repairs.
- FRT, indicates that brake application can be modulated, but the release will be complete (direct release).
- PASS, indicates that the brake application and release can be modulated by the operator (graduated release).

Choice of operating conditions will be determined by Railway Company Regulations and Directives.

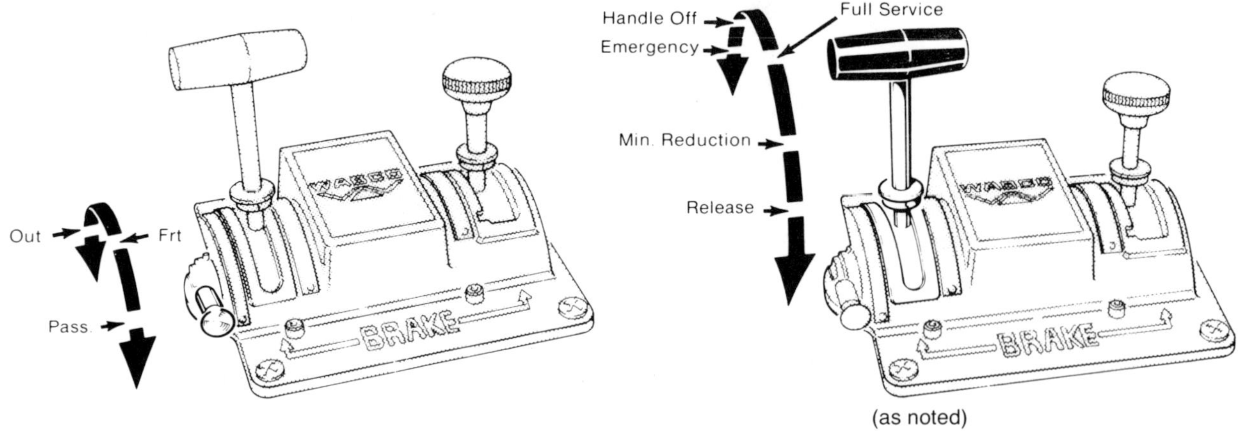

Fig. 2-13 Brake Cut-Off Valve

Fig. 2-14 Automatic Brake

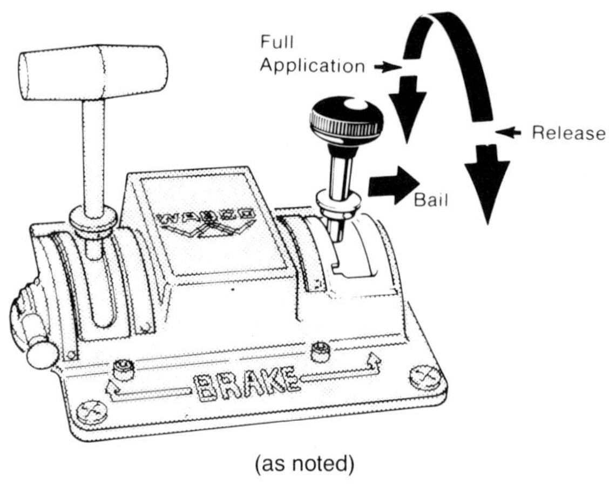

Fig. 2-15 Independent Brake

Automatic Brake Lever (Fig. 2-14)
Movement of this handle through detent provides positive location for the following operation modes: "Release", "Minimum Reduction", "Suppression", "Full Service", "Handle Off" and "Emergency". As indicated above by depressing the lock ring, this handle can be removed when in the HANDLE OFF position.

- RELEASE. In this position the brakes are completely released.
- MINIMUM REDUCTION. Movement of the automatic brake handle to this position initiates a 37 to 48 kPa (5½ to 7 psi) pressure reduction in the equalizing reservoir. This reduction creates a similar reduction in brake pressure by means of the 30-CW module. As the handle is moved farther into the service zone beyond "minimum reduction" a regulating cam initiates pressure reduction in the brake valve module and ultimately in the brake pipe.
- SERVICE. From the minimum reduction position to the full service application, the brake action can be modulated from minimum to maximum by gradually moving the lever forward.
- FULL SERVICE. This position is similar to the "suppression" one except for regulating an approximate 166 to 179 kPa (24-26 psi) brake pipe pressure reduction.
- SUPPRESSION. Located in the service zone, the "suppression" position calls for an approximate 120 kPa (17 psi) brake pipe reduction and will also nullify an impending penalty brake application when the handle is moved in this position before expiration of four to six seconds after warning of an impending penalty brake application has been sounded.
- HANDLE OFF. Depending on the amount of time the handle remains in this position, brake pipe pressure may be reduced down to 55 to 69 kPa (8-10 psi). This position is used when the locomotive is used in trail service. The locking ring will release the handle when in this position.
- EMERGENCY. In this position the brake pipe will be vented to zero pressure at the fastest rate possible. This is an emergency brake application and signals to initiate dynamic cut-out and power cut-out will be generated in this position.

Independent Brake Valve (Fig. 2-15)
This valve provides means of controlling the locomotive brake cylinder pressure proportionally to the lever movement independently from the brake system of the train. This valve has two detented positions: "Release" and "Application".

- RELEASE. In this position the locomotive brakes are completely released. The handle can be removed by depressing the locking ring.
- FULL APPLICATION. When the handle is moved from the release position, brake application starts. This application is directly proportional to the degree of movement from the released position. At the "full application" position the locomotive brakes are fully applied.
- BAIL. When in the release position, by moving the handle to the right, the locomotive brakes will be released, while an automatic brake application is in effect throughout the train.

(17) **Horn**
The horn control is situated at the edge of the desk directly in front of the operator. Pulling the control will actuate the air horn.

(18) **Brake Pipe Pressure Regulator**
The brake pressure control removable knob can be introduced in the air valve module at the edge of the horizontal panel of the console, directly in front of the brake valve. With the automatic brake valve handle in the release position, this valve will bring the brake pipe to the desired pressure. The automatic brake valve will maintain this selected pressure against overcharge or leakage.

SIDE VERTICAL PANEL (Fig. 2-16)
This panel is situated at the left of the operator, directly over the console desk surface. It contains the following groups of controls: a group of switches for controlling the power circuits, a digital speed indicator, a set of indicator lights and miscellaneous controls.

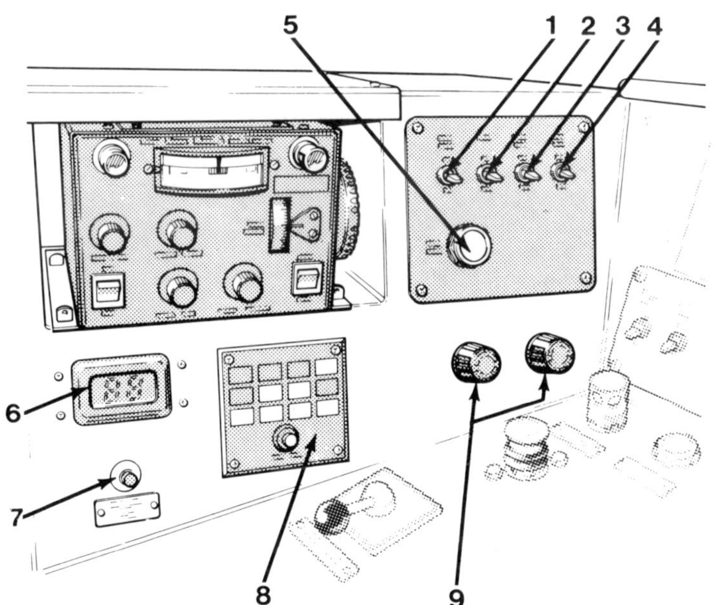

1. Thyristor Control
2. Control
3. Power Limiter
4. Engine Run
5. Reset Push Button
6. Digital Speed Indicator
7. Attendent Call
8. Fault Indicator Lights
9. Windshield Wipers

Fig. 2-16 Side Vertical Panel

(1) **Thyristor Control**
This toggle switch is the first on the left of the switch group at the top of the side panel. Push up to the ON position to complete the power circuits to the traction motors.

(2) **Control**
The Control Switch provides power to the various low voltage control circuits.

(3) **Power Limiter**
When this switch is in the ON position the power of the locomotive is reduced by blocking bridge 2. This function can be used to reduce the power demand or improve the power factor, at certain speeds.

(4) **Engine Run**
This switch should normally be OFF. Only if a diesel locomotive is to operate in the multiple unit consist, then this switch must be in the ON position to provide control of the diesel power system from the leading electric locomotive. Note that a diesel locomotive can be used in the trailing position only when operating with the GF6C electric locomotives.

(5) **Reset Push Button**
When a fault condition is indicated by a warning light on any of the indicating light panels, bring the throttle/brake lever to the "O" position and then press the reset button. This will reset the system and the locomotive will resume operation. If the fault has not been cleared by this procedure, the affected circuit may have to be isolated before attempting to reset again. Refer to the "Unusual Operating Conditions" section in this manual for more detailed instructions in fault reset procedures.

(6) **Digital Speed Indicator**
This digital speed indicator is situated at the left of the side vertical panel. It indicates the speed in km/h as read by the locomotive wheels. Note that a slight difference of reading between the digital and analog meters is possible due to the difference in detection mode. At times the digital reading might be slightly higher than the radar reading due to wheel slip.

(7) **Attendant Call**
Situated just below the digital speed indicator, this button when pushed in any unit of the consist will cause the alarm to ring in all the units.

(8) **Fault Indicator Lights** (Fig. 2-17)
The following warning lights are situated in the middle of this side panel. For corrective measures, refer to the "UNUSUAL OPERATING CONDITIONS" section in this manual.

(8A) MAIN C.B. OPEN (Circuit Breaker) – RED
Indicates the main breaker has tripped for some reason.

(8B) OVERCURRENT BRAKE – RED
Indicates an overcurrent condition in the traction motor armatures during dynamic braking.

(8C) PCS OPEN (Pressure Control Switch) – RED
Indicates that a penalty or emergency brake application has occured.

(8D) PWR DEMAND LIMIT (Power) – WHITE
Power limit switch is in the ON position.

(8E) APL NO PWR. (Announcer Panel no Power) – RED
Indicates any of the following:

1. High air temperature in the convertor
2. Low oil pressure in the convertor
3. Low oil pressure in the transformer
4. High oil temperature in the transformer
5. Phase fault in the convertor
6. Auxiliary convertor is disabled.

(8F) SAND – WHITE
Indicates that manual sanding is in operation.

(8G) WHEEL SLIP – WHITE
Intermittent flashing of WHEEL SLIP lights indicates moderate to severe wheel slip. Continuous WHEEL SLIP light could indicate a locked wheel condition.

(8H) LOCK. WHEEL (Locked Wheel) – RED
Indicates a locked wheel condition and/or a failed traction motor pickup.

NOTE: See "Section 4" for detailed instructions.

The push button below this warning light panel is a lamp test. Prior to starting operations, the lamps should be tested. All bulbs that do not light during this test should be changed before starting.

(9) **Windshield Wipers**
Two control knobs below the switches control the three windshield wipers. The knob towards the front of the locomotive controls the wiper in front of the operator, while the knob towards the rear of the unit controls the wiper on the center window. The wipers are operated by compressed air and their speed can be adjusted as required. Turn the knobs completely clockwise to stop the wipers. Turn them progressively counterclockwise to operate them faster.

LOWER SECTION (Fig. 2-18)
Some controls are situated below the main level of the console. They are the following:

(19) **Two-Way Radio**
This radio system is situated directly on the left side of the locomotive operator. For operation of this radio refer to the radio operating instruction manual. The hand controlled microphone hangs above the radio.

(20) **Gauge Light Dimmer**
It is situated just below the radio and to the left of the console. It will control all the gauge lights on the console. Another dimmer control is situated on the overhead panel to dim those lights.

(21) **Cab Heater Controls**
The two four position switches, just below the two-way radio, control the cab heaters on the operator's side; the one on the left controls the heat supplied directly under the control console and has four positions:

- OFF
- LO
- HI
- FAN

The one on the right of the panel controls the auxiliary sidewall heater and has three positions:

- OFF
- LO
- HI

Multiple Unit Control Valve
The multiple unit (MU-2) valve is located on the lower section of this side panel. The purpose of this valve is to select the mode of operation of the brake system according to whether the locomotive is leading or trailing. This valve has two positions:

- LEAD or DEAD
- TRAILING

The valve position is selected by pushing the knob in and turning it to the desired position.

(22) **Safety Control Foot Pedal**
This pedal situated directly below the control console desk must be depressed at all times during the operation of the locomotive and/or if brake cylinder pressure is below 172 kPa (25 psi).

Failure to keep this foot pedal depressed with brake cylinder pressure less than 172 kPa (25 psi), will cause a signal whistle to sound for 4 to 6 seconds. If no corrective action is taken a penalty application of the brakes will occur and power to the traction motors will be cut.

OVERHEAD CONTROLS
A series of instruments, gauges and switches are situated directly over the operator's control console. There are four sub-panels on the operator's side and one on the assistant's side.

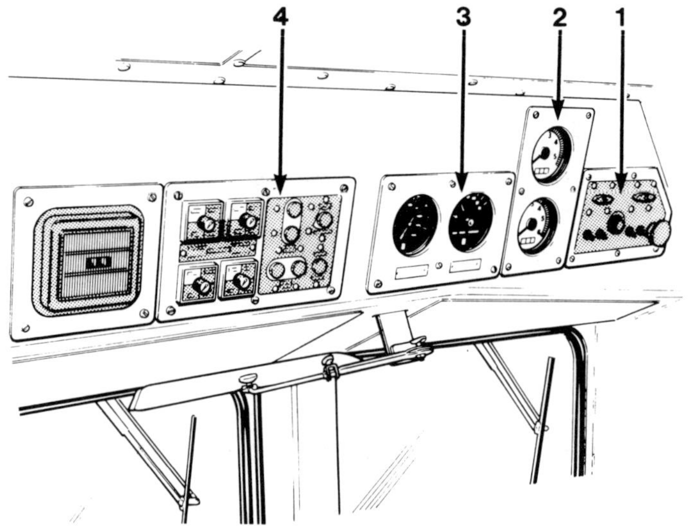

1. Switch Panel
2. Electrical Panel
3. Main Brake Panel
4. Classification and Indication Light Panel

Fig. 2-19 Overhead Controls

OVERHEAD PANELS - OPERATOR'S SIDE (Fig. 2-19)
The following sub-panel and gauges are situated at the operator's side:

Switch Panel (Fig. 2-20)

(1) FRONT AND REAR HEADLIGHTS SELECTOR SWITCH
These two (2) four-position switches are situated at the top of the first right overhead panel. They can be placed as required in the four (4) following positions:

- OFF
- DIMMED
- MEDIUM
- BRIGHT

(2) WINDSHIELD HEATER
This toggle switch activates the windshield heaters when required to disperse mist or frost.

(3) MIRROR HEATER
Same as above but for the protection of the rear view mirrors.

(4) GAUGE LIGHT DIMMER
This small rheostat will control the brightness of the gauges on the overhead panels. Turning it clockwise will increase the light, counterclockwise will dim it. Note that the operator might require a different level of gauge illumination for the instruments on the console than those overhead due to the difference in distance from him.

(5) PANTOGRAPH
Prior to operation, the normal position for this switch is DOWN. After all the selector switches and controls of the individual locomotives of the MU consist have been positioned, placing this switch in the UP position will raise all the selected pantographs of the consist. Returning this switch to the down position will lower all the pantographs in the consist simultaneously.

(6) CIRCUIT BREAKER
This toggle switch operates the main circuit breaker. Prior to operation, the normal position for this switch is OPEN. Once the pantographs have been raised, the circuit breaker can then be placed in the CLOSED position.

(7) EMERGENCY POWER OFF
This large red push button at the extreme right side of the overhead panel, when pressed, will cut the power by opening the main circuit breaker and will lower all the pantographs of the consist.

CAUTION: THIS BUTTON DOES NOT APPLY THE AIR BRAKES.

Electrical Panel (Fig. 2-21)

(1) POWER INDICATOR
The top dial on the second sub-panel indicates the total power consumption of the locomotive at any given moment. This information is to be used when there are power restrictions on the line.

(2) LINE VOLTAGE
This voltmeter indicates at all times the voltage on the catenary system. This indication will warn the operator of a voltage drop that might require power reduction on the locomotives.

Main Brake Panel (Fig. 2-22)

(1) AIR FLOW INDICATOR
This gauge indicates at all times the actual air flow in the brake pipe. This gauge will give an indication of the performance of the air brake system and provide a warning of abnormal air losses.

(2) SIGNAL AIR PRESSURE GAUGE
This gauge indicates the pressure on the Signal Air Pipe system of communication from train to locomotive. A whistle will sound in the cab of the locomotive when the system is activated. Pressure indicated should be in accordance with the specifications of the Railway Company.

Classification and Indication Lights Panel (Fig. 2-23)

(1) FRONT HEADLIGHT INDICATOR LIGHTS
On the left side of the overhead panels, operator's side, there is a small group of six (6) pilot lights. At the top left corner of this group are two (2) indicator lights placed one above the other. Failure of one of them to light will indicate that the corresponding headlight has failed and must be replaced at the earliest opportunity, following the Company's regulations.

(2) REAR HEADLIGHT INDICATOR LIGHTS
These two (2) horizontal pilot lights are situated directly below those for the front headlights. Their operation is the same as above.

(3) WINDSHIELD HEATER INDICATOR LIGHT
This pilot light at the right of the two vertical ones will indicate if the defrosting heater is operating.

(4) MIRROR HEATER INDICATOR LIGHT
This light at the lower right hand corner is the same as above but for the rear view mirrors.

(5) CLASSIFICATION LIGHT SWITCHES
These switches control the GREEN, RED, WHITE classification lights. Set these in conformity with the Railway Company's regulations.

Overhead Panel – Assistant's Side
Only two sets of switches are situated on the assistant's side of the overhead controls. These are:

- Windshield Wipers
- Reading Light

WINDSHIELD WIPER CONTROL
This is the same type of control as on the operator's vertical side panel. This knob controls the wiper on the assistant's side window. Turn it completely clockwise to stop it, counterclockwise to start it and speed it up.

READING LIGHT
This toggle switch will switch on the reading light on the assistant's side.

Miscellaneous Controls
The following controls should be noted in the control cab and in the short hood:

- Heater Control Assistant Side
- Hot Plate Control
- Hand Brake

HEATER CONTROL ASSISTANT SIDE

Situated on the heater in front of the assistant's seat, this rotary switch has three positions:

- OFF
- LO
- HI

HOT PLATE CONTROL

Situated on the hot plate frame, this rotary switch controls the hot plate element.

HAND BRAKE

The hand brake is situated in the access passage through the short hood. Turning the circular wheel all the way clockwise will release the brakes; turning it counterclockwise will apply the brakes. During a layover period always apply this brake. Follow the Company's regulations as to all safety procedures for ensuring a safe layover.

SECTION 3
OPERATION

INTRODUCTION
This section of the manual covers recommended procedures for operation of the GF6C Electric Locomotive. The procedures are outlined briefly and do not contain detailed explanations of equipment location or function.

The information in this section is arranged in sequence, commencing with inspections in preparation for service, handling a light locomotive, coupling to train, and routine operating phases. The various operating situations and special features such as dynamic braking are also covered.

PREPARATION FOR SERVICE
Preparation for service includes ground and roof inspection, machine room inspection, cab inspection, and set up procedures for a single unit or multiple unit operation.

GROUND AND ROOF INSPECTION
Ground and roof inspection includes general visual inspection and proper placement of controls as specified below:

1. Inspect for air or oil leaks.
2. Inspect for loose and dragging equipment.
3. Inspect for proper hose and jumper connection if in multiple unit.
4. Inspect for proper position of all angle cocks and shut-off valves.
5. Inspect for satisfactory condition of all brake shoes.
6. Ensure that the brake cylinder valve is cut-in for all cylinders.
7. Check the humidity indicators for the compressed air dryer situated on the left side of the locomotive under frame.
8. Inspect the roof-mounted equipment from the ground level for the condition of the bus bars and of the pantograph contact shoes.

MACHINE ROOM INSPECTION
Machine room inspection includes a general visual inspection to ensure that all is in order, panels closed, no sign of oil leaks, and no loose equipment has been left behind. This inspection also includes the following:

1. Check air compressor oil level and temperature, item 1.
2. Check transformer and thyristor convertor oil level, item 2.
3. Check the air dryer indicators on the compressed air equipment, item 3.
4. Check the air dryer indicators on the transformer, item 4.
5. Check for air or oil leaks.

Pantograph Selection (Fig. 2-1)
In each machine room, if in multiple unit operation, the chosen pantograph must be set and the auxiliary compressor started prior to leaving each machine room as outlined below.

1. Check the Grounding Switch and make sure it is in the UNGROUNDED position.

GROUNDING SWITCH OPERATION (Fig. 3-1)

To ground the locomotive:

a. Lower both pantographs.
b. Turn the pantograph selector switch to the OFF position.
c. Turn Isolating Cock to the ISOLATED position.

d. Turn the Interlock to the GROUNDED position.
e. Turn the main Grounding Switch handle to the GROUNDED position and latch it in position.

To unground the locomotive:

a. Release the latch and return the main Grounding Switch handle to the UNGROUNDED position.
b. Return the Interlock to the UNGROUNDED position.
c. Return the Isolating Cock to the NORMAL position.

2. Set the pneumatic valve to the required position, cab end or long hood end.
3. Set the electrical selector switch, from the OFF position to the corresponding position.
4. Press the starter button for the auxiliary air compressor.

Note: If the grounding switch has remained in the GROUNDED position, the pantograph cannot be raised and the Main Circuit Breaker cannot be closed.

Once the pantograph has been selected, the machine room can be left and the next machine room inspected. Finally when all the pantographs have been selected and the operator is in the controlling cab, he will be able to raise all the selected pantographs with the one pantograph "UP-DOWN" switch when ready to do so.

Summary of Machine Room Controls

Grounding Switch .... UNGROUNDED
Pantograph
  Electrical Selector .... SELECTED
  Pneumatic Selector .... SELECTED
  Auxiliary Compressor .... ON

TRAILING CAB INSPECTION

Safety Note: All jumpers and brake hoses must be set before further action is taken.

After a general inspection to ensure that the conditions are normal, the following points must be followed to set the locomotive for trailing duty:

Note: The pantograph "UP-DOWN" switch MUST REMAIN in the DOWN position on all locomotives. Only when the consist is ready for operation, all selected pantographs will be raised from the leading cab.

1. Place all the breakers in the black areas of the breaker panels in the ON position. These are the breakers marked with an (★) asterisk in the descriptive sections of this manual.
2. Place the headlight switch on the Locomotive Control Panel at the required position depending on the position of the locomotive in the consist.
3. The throttle lever must be left in the "O" position.
4. If it has not yet been done, remove the Automatic Brake lever and the Independent Brake lever.
5. Place the Reverser handle in the NEUTRAL position and remove it. This will electrically interlock the main controls.
6. Place the Cut-off Valve in the OUT position and the MU valve in the TRAILING position.
7. Place the Engine Run, Power Limiter, Thyristor Control and the Control switch in the OFF position. These functions will be controlled from the leading cab.
8. Other switches and breakers are to be set as required by operating conditions and Company regulations.
9. Finally, before leaving the cab, place the Main Circuit Breaker switch in the CLOSED position.

Summary of Trailing Cab Controls

Breakers
- All black breakers (★) .... ON
- Other breakers .... AS REQUIRED

Battery switch .... ON
Ground relay cut-out switch .... ON
Control switch .... OFF
Isolation switch .... RUN
Engine run switch .... OFF
Power Limiter switch .... OFF
Thyristor Control switch .... OFF
Main Circuit Breaker switch .... CLOSED
Pantograph switch .... DOWN
Throttle .... O
Reverse handle .... NEUTRAL & REMOVED

Brakes
- Automatic brake handle .... REMOVED
- Independent brake handle .... REMOVED
- Cut-off valve .... OUT
- MU Valve .... TRAILING

LEAD CAB INSPECTION

Safety Note: All jumpers and brake hoses must be connected before further action is taken.

Once all the trailing units of the consist have been set as above, all operations are to be transferred to the lead unit. Follow the instructions given below and all the rules and regulations of the Railway Company.

1. Place all the breakers in the black areas of the breaker panels in the ON position. These are the breakers marked with an (★) asterisk in the descriptive section of this manual.
2. Place the headlight switch on the Locomotive Control Panel in the lead position.
3. The throttle lever must be left in the "O" position.
4. Place the two brake handles in position, the Automatic brake in RELEASE, the independent brake in full APPLICATION. Since the APL system is not yet energized and therefore the main air compressor has not yet started, brake application may be delayed due to lack of compressed air in the main reservoirs.
5. Place the Reverser handle in position and leave it in NEUTRAL until such time as the locomotive will be ready to operate.
6. Place the Cut-off valve in the PASS or FRT position and the MU valve in the LEAD position.
7. Place the Control switch in the ON position. Normally the Engine Run switch will be placed in the OFF position unless an operational diesel locomotive is included in the consist.
8. Place the Power Limiter in the OFF position unless specifically instructed to the contrary by Operation Control.
9. Place the Thyristor Control switch in the OFF position, and place the Isolation switch in the STANDBY position.
10. Place the Pantographs switch in the UP position. All the selected pantographs of the consist will now be raised.
11. Place the Main Circuit Breaker in the CLOSED position. The APL system is now in operation, the main compressor is supplying the brake system and the independent brake cylinder pressure is now being applied.

Summary of Lead Cab Controls

Breakers
- All black breakers (★) . . . . ON
- Other breakers . . . . AS REQUIRED

Battery switch . . . . ON
Locomotive control panel . . . . AS REQUIRED
Ground relay cut-out switch . . . . ON
Control switch . . . . ON
Isolation switch . . . . STANDBY
Power Limiter Switch . . . . OFF
Thyristor Control switch . . . . OFF
Main Circuit Breaker switch . . . . CLOSED
Pantograph switch . . . . UP
Throttle . . . . O
Engine Run (unless diesel in MU) . . . . OFF
Reverser handle . . . . NEUTRAL
Brakes
- Automatic brake . . . . RELEASED
- Independent brake . . . . APPLIED
- Cut-off Valve . . . . FRT or PASS.
- MU valve . . . . LEADING

DRAINING AIR RESERVOIRS AND STRAINERS

The air reservoirs or filters should be drained periodically whether or not the equipment is provided with an automatic drain valve. Follow the applicable maintenance schedule.

STARTING LIGHT OR IN A CONSIST

Having complied with all rules and regulations of the Railway Company and ready to move the locomotive or the consist, proceed as follows:

1. Make sure the main reservoir air pressure is NORMAL, from 898 to 965 kPa (130 to 140 psi).
2. Check for proper application and release of air brakes.
3. Release hand brake and remove any blocking from under the wheels.
4. Switch on the headlights as required by operating conditions.
5. Place the Isolation switch in the RUN position and the Thyristor switch in the ON position.
6. Place the Reverser handle in the required direction of travel, and slowly RELEASE the Independent Brake.
7. Open the throttle to RUN 1, 2 or 3, as needed to move the locomotive to the desired speed.
8. Throttle should be in the "O" position before coming to a complete stop.
9. Reverser handle should be moved to change position ONLY when locomotive is completely stopped.

Summary of Starting Controls

Breakers
- All black breakers (★) . . . . ON
- Other breakers . . . . AS REQUIRED

Battery switch . . . . ON
Headlights . . . . ON
Locomotive control panel . . . . AS REQUIRED
Ground relay cut-out switch . . . . ON
Control switch . . . . ON
Isolation switch . . . . RUN
Power Limiter switch . . . . OFF
Thyristor Control switch . . . . ON

Main Circuit Breaker switch . . . . CLOSED
Pantograph switch . . . . UP
Engine Run (unless diesel in MU) switch . . . . OFF
Dead Man pedal . . . . DOWN
Reverser handle . . . . AS REQUIRED
Throttle . . . . AS REQUIRED
Brakes
  Automatic Brake . . . . RELEASED
  Independent brake . . . . RELEASED
  Cut-off valve . . . . PASS or FRT.
  MU valve . . . . LEADING

COUPLING LOCOMOTIVE UNITS TOGETHER

When coupling units together for multiple unit operation, the procedures below should be followed:

1. Couple and stretch units to ensure the couplers are well locked.
2. DROP all pantographs on the consist.

   Note: Always lower all pantographs of a consist prior to setting or removing jumper cables between locomotives of a consist.
3. Apply all the required jumper cables and brake hoses between the locomotives of the consist. See Fig. 3-2.
4. Open air hose cutout cocks as required on all units.
5. Attach platform safety chains between units.
6. Proceed with the machine room inspection as described above.
7. Proceed with the trailing cab inspection and setting as described above, see page 3-4.

COUPLING LOCOMOTIVE TO TRAIN

Locomotives must be coupled to the train using the same care taken when coupling the units together. After coupling, make the following checks:

1. Check that the couplers are locked by stretching the connections.
2. Connect the air hoses.
3. Slowly open air valve on the locomotive and train to cut in brakes.
4. Pump up air if necessary using the following procedures.

Pumping Up Air

It is possible that after cutting in the train air brakes the main reservoir air gauge may fall below the trainline pressure. On the electric locomotive the air compressor will automatically compensate and provide maximum air output, regardless of throttle position.

Brake Pipe Leakage Test

Prior to moving the train, a leakage test must be performed. This is accomplished in the following manner:

1. The cut-off valve is positioned in the PASS or FRT position as set out in the Railway Company's regulations.
2. Move the automatic brake valve handle gradually into service position until the equalizing reservoir gauge indicates that a 1.05 kg/cm^2 (15 psi) reduction has been made.
3. Without any further movement of the automatic brake valve handle, observe the brake pipe gauge until this pressure has dropped 1.05 kg/cm^2 (15 psi) and exhaust has stopped blowing.
4. At this moment place the cut-off valve lever in the OUT position. This will cut out the maintaining function of the brake valve.

5. From the instant the cut-off valve is placed in the OUT position, the brake pipe gauge should be observed and any possible drop in the brake pipe pressure should be timed for one minute. Brake pipe leakage must not exceed the rate established by the railroad rules.
6. After checking trainline leakage for one minute and the results are observed to be within required limits, return the cut-off valve lever to the PASS or FRT position and proceed to reduce the equalizing gauge pressure until the pressure is the same as the brake pipe gauge pressure. This is done by moving the automatic brake valve handle gradually until full application of the brakes has been obtained.
7. After the pipe leakage test has been completed, return the automatic brake valve to the release position.

RUNNING PROCEDURES

This section will deal with the main points of locomotive operation under normal conditions. The information given here is only a guide and at all times the operating rules and regulations of the Railway Company will have precedence and must be followed.

STARTING A TRAIN

The method to be used in starting a train will depend upon many factors, such as number and type of locomotives in the consist, weight and length of the train and the amount of slack in the train. Also the weather, grade and track conditions must be taken into account. Since all these factors are variable, specific train starting instructions cannot be given, and it will be left to the good judgement of the operator to apply the required amount of power to suit the requirements. There are, however, certain general considerations that should be observed. They are discussed in the following paragraphs.

On an electric locomotive the tractive effort at start can be very high and will be applied at a controlled rate to the wheels the moment the throttle is advanced. This makes it imperative that the air brakes be completely released before attempting to start the train.

The locomotive has sufficiently high tractive effort to allow it to start most trains without taking slack. At times, however, it might be necessary to take slack to start a train. Care should be taken in such cases to prevent excessive locomotive acceleration to avoid undue shock to the draft gear, couplers and lading.

Proper throttle handling is important when starting trains, since it has a direct bearing on the power developed. As the throttle is advanced, a power increase occurs immediately, and the power applied is proportional to the throttle position which should be advanced one position at a time. Start the train in the lowest possible throttle position to keep the locomotive speed as low as possible until all slack has been taken and the train fully stretched. Sometimes it is advisable to reduce the throttle a position or two when the locomotive starts moving to prevent taking up the slack too quickly.

Summary

When ready to start, the following procedures are recommended:

1. Move the reverser handle to the desired direction.
2. Release both automatic and independent air brakes.
3. Open the throttle one position at a time as follows:
   a. TO No. 1 – The locomotive will immediately begin to produce tractive effort which may be sufficient to start the train.

      Note: The locomotive power control system makes it generally unnecessary to apply the locomotive independent brake or to manipulate the throttle between No. 1 and 0 during starting procedures.

   b. TO No. 2, 3 OR HIGHER – Experience and the demands of the schedule will determine the best procedure until the train moves.
4. Reduce throttle one or more positions if acceleration is too rapid.
5. After the train is stretched, advance throttle as desired.

Note: During starting procedures, manual sanding is available up to 8 km/h (5 mph), at which point it will be automatically disabled, and replaced by automatic sanding as dictated by the wheel creep control system.

ACCELERATING A TRAIN

Once the train has started, the throttle can be advanced as rapidly as required for accelerating the train. The rate of advance of the throttle depends on the train and exterior conditions. Advancing the throttle one position at a time will reduce tendency of wheel slippage.

The ammeters will give the operator the best indication for throttle operation during acceleration. When moving the throttle up one position, the ammeters will rise, and as energy is absorbed by the train, they will drop back to some degree. When all have dropped back, the throttle can be advanced one more position, until position 8 is reached.

AIR BRAKING WITH POWER

The air brake handling procedures are left to the discretion of the Railroad Company. However, it must be remembered that for any throttle position, the draw bar pull rapidly increases as the speed decreases. This pull might become great enough to part the train unless the throttle is also reduced as the speed drops. Since the pull of the locomotive is indicated by the amperage on the ammeters, the operator can maintain a constant pull on the train during a slowdown, by keeping a steady amperage on the meter. This is done by reducing the throttle by one position each time the ammeters start to rise.

During a power brake, it is recommended that the independent brake be maintained in the RELEASED position. The throttle MUST BE IN THE 0 POSITION before the locomotive comes to a complete stop.

POWER STALL

NEVER hold the train at a standstill on a grade or with the brake applied and the throttle open for power. Extensive damage will result to the traction motors.

OPERATING OVER RAIL CROSSING

When operating the locomotive over 40 km/h (25 mph), reduce the throttle position 4 at least eight seconds before the locomotive reaches a rail crossing. If the locomotive is running at position 4, or with a speed less than 40 km/h (25 mph), allow the same eight seconds with the throttle in the next lower position. Advance the throttle after all the units of the consist have passed over the crossing. This procedure is necessary to ensure lower electrical and mechanical levels of effort before the shocks that occur at all rail crossings reach the motor brushes.

RUNNING THROUGH WATER

Under ABSOLUTELY NO CIRCUMSTANCES should the locomotive be operated through water deep enough to reach the underside of the traction motors. Water deeper than 75 mm (3″) above the rail is likely to cause traction motor damage.

When passing through any water on the rails, exercise every precaution under the circumstances, traveling very slowly, never exceeding 3 to 5 km/h (2 to 3 mph).

LOCKED WHEEL LIGHT

If the "Locked Wheel" light on the side panel of the console comes on and burns continuously, accompanied by the alarm bell, a locked wheel condition exists. Follow the Railway procedures for corrective action.

DYNAMIC BRAKING

Dynamic braking is very effective in slowing down the train speed, especially while descending grades, thus reducing the need for using air brakes.

The maximum continuous braking effort available is 264 kN (59,350 lb). This braking effort is achieved at approximately 48 km/h (29.6 mph) and is maintained down to a speed of approximately 9 km/h (5.6 mph) by means of the "Extended Range" system. At higher and lower speeds the effectiveness decreases. For this reason it is important that dynamic braking be started BEFORE the train speed becomes excessive, and during dynamic braking the application of air brakes should be continuously monitored to maintain the speed of the train within the optimum range of the dynamic brake.

Within this range of optimum braking efficiency small increments of braking effect will be felt on the train; this is not cause for alarm and brakes need not be adjusted.

To operate the dynamic brakes, proceed as follows:

1. Return the throttle to the 0 position and maintain it in this position for a sufficient length of time to allow the slack to run in and ensure proper train handling before further movement.
2. Lift the throttle/brake over the stop gate and move the handle slightly into the dynamic braking zone. A slight braking effect will occur as shown on the ammeter.
3. After the slack has been bunched, braking effort may be increased by slowly advancing the handle to the maximum braking position. Dependent on train speed, maximum braking current limited to 760 amperes can be achieved over a range of Dynamic brake handle positions.
4. When traction motor armature current exceeds a predetermined value, the "Overcurrent Brake" light on the console of the affected locomotive will come on. The fault reset switch on the affected locomotive must be depressed to restore dynamic braking on this locomotive. If overcurrent indications are repeated the locomotive must be taken out of dynamic braking service by placing the Dynamic Brake Cut-out switch on the Locomotive Control Panel in the ON position.
5. When an open circuit is detected in the grid circuit, dynamic braking on this unit will be removed. The OCP reset switch on the affected locomotive must be depressed to restore dynamic braking on this unit. If "Open Circuit" indications are repeated the locomotive must be taken out of dynamic braking service by placing the Dynamic Brake Cut-out switch on the Locomotive Control Panel in the ON position.
6. Since the total braking effort will be reduced when a unit in a consist has had the dynamic cut out, it may be necessary to use the automatic brake in conjunction with the dynamic brake. However, the independent brake must be KEPT FULLY RELEASED whenever the dynamic brake is applied or the wheels may slide. As the speed decreases below 16 km/h (10 mph) the dynamic brake becomes less effective. When the speed further decreases, the dynamic brake can be completely released by placing the throttle/brake handle back in the 0 position, past the gate, and applying the independent brake at the same time to prevent the slack from running out.

DYNAMIC BRAKE WHEEL SLIP/SLIDE CONTROL

The interconnection arrangement of the resistor grid and the motors is such that an automatic wheel slip/slide control will take place. This and the braking adjustment achieved by automatic switching of the resistor grid elements, result in a relatively smooth and continuous dynamic brake operation. Automatic sanding may occur during poor adhesion conditions, but manual sanding is available to the operator.

OPERATION IN HELPER SERVICE

Basically, there is no difference in the instructions for operating the locomotive as a helper or with a helper. In most instances it is desirable to get over the grade in the shortest possible time. Thus, wherever possible, operation on the grade should be at full throttle. Throttle can be reduced, however, when wheel slip causes lurching that may threaten to break the train. During full throttle operation on grades, the traction motor temperature simulator may reduce the power to avoid overheating. This will take place automatically and the throttle position need not be changed.

ISOLATING A UNIT

When it becomes necessary to isolate a locomotive, proceed as follows:

1. When operating under power or in dynamic brake mode, a unit may be isolated at any time, but discretion as to timing and urgency should be used.
   a. Bring back the throttle to the "0" position.
   b. Place the Isolation Switch on the Locomotive Control panel in the STAND-BY position. Also place the Thyristor Control Switch on the console in the OFF position. This will isolate the unit. Other locomotives in the consist will operate normally.
2. If a complete shutdown is required, the locomotive will have to be towed "dead" in the train.

TOWING A LOCOMOTIVE DEAD IN A TRAIN

1. Open the compressor breaker and reduce main reservoir pressure to below brake pipe pressure.
2. Place Cut-off valve in the OUT position.
3. Place the Multiple Unit Valve in the DEAD position.
4. Place the Automatic brake handle in the HANDLE OFF position and remove it.
5. Place the Independent Brake handle in the RELEASE position and remove handle.
6. Place the Main Circuit Breaker Switch in the OPEN position and the Pantograph switch in the DOWN position.
7. Place the Grounding Switch in the machine room in the GROUNDED position.
8. Place all the circuit breakers in the OFF position.
9. Place all control switches in the OFF position.
10. Place the battery rotary switch in the OFF position.

LEAVING THE LOCOMOTIVE UNATTENDED (PANTOGRAPH UP)

If at any time it is necessary to leave the locomotive unattended with auxiliary systems in operation, the following procedures should be followed:

1. Observe all rules and regulations of the Railway Company which may apply and which may include the locking of the cab doors.
2. Make a complete brake application.
3. Place the Isolation Switch in the STAND-BY position.
4. Place the throttle in the 0 position, the reverser handle in the NEUTRAL position and remove it to prevent accidental operation of the locomotive.

LEAVING THE LOCOMOTIVE UNATTENDED (PANTOGRAPH DOWN)

When the locomotive is to be left unattended with the auxiliary systems OFF and the pantograph down, the following procedures should be followed:

1. Observe all rules and regulations of the Railway Company which may apply and which may include the locking of the cab doors.
2. Make a complete brake application.
3. Place the Isolation Switch in the STOP position.
4. Place the Main Breaker Switch in the OPEN position and the Pantograph switch in the DOWN position.
5. Open all breakers or as required.
6. Place the throttle in the 0 position, the reverser handle in the NEUTRAL position and remove it to prevent any accidental operation of the locomotive.
7. Open battery switch.

LEAVING THE LOCOMOTIVE UNATTENDED (PANTOGRAPH DOWN AND CONNECTED TO SHORELINE POWER)

When the locomotive is to be left unattended with the pantograph down and connected to shoreline power, the following procedure must be followed:

1. Observe all rules and regulations of the Railway Company which may apply and which may include the locking of the cab doors.
2. Make a complete brake application.
3. Place the Isolation Switch in the STAND-BY position.

4. 480 VAC:
   If 480 VAC is required to operate the auxiliary motors, the load capability of the power source will have to be established by the Railway, to enable adequate operation of any or all of the motors.
   a. Before inserting the layover plug into either 480 V receptacle on the locomotive, ensure the required 480 VAC circuit breakers are switched ON.
   b. Ensure the LAYOVER circuit breaker is closed.
   c. Move the LAYOVER switch on the locomotive control panel to the ON position.
5. 240 VAC:
   a. Insert the layover plug into either 240 V receptacle on the locomotive.
   b. Ensure the LAYOVER circuit breaker is closed.
   c. Place the required circuit breakers (240 VAC or 120 VAC) in the ON position.
   d. Move the LAYOVER switch on the locomotive control panel to the ON position.

POWER LIMITER OPERATION

This locomotive is equipped with a control limiting the consumption of supply power. The Railroad Operation will dictate when the trains will operate in this reduced power consumption mode.

When the operator is instructed to reduce power demand, the "Power Limit" switch on the control console must be placed in the ON position. This action effectively limits the power consumption by 50%. However, operation is not affected during start-up of the train; only within a given condition, the maximum speed will be reduced.

SECTION 4
UNUSUAL OPERATING CONDITIONS

INTRODUCTION

This section covers operational problems that may occur on the road and suggests remedial actions that may be taken by the operator in response to the trouble.

Safety devices automatically protect the equipment in case of faulty operation of most components. In general this protection is obtained by one of the following methods:

1. Shutdown of the auxiliary power supply function or disabling of a function such as dynamic braking. In some instances manual resetting of the function may be necessary or automatic resetting after a time delay may be provided.
2. Power output reduction with corresponding speed reduction.
3. Back-up regulation for the protection of the equipment.

INDICATING LIGHT PANEL - CONTROL CONSOLE (Fig. 2-16)

The following conventions are applied to the following information:

References – In the following text all reference numbers pertain to the illustration 2-17.

Indentations — In the following text the indentations have the following meaning:

> **Margin** text will describe the CAUSE of a given problem.
>
> > **First Indent** will list all the EFFECTS that can be observed by the operator.
> >
> > > **Second Indent** will list all the corrective ACTIONS that must be taken by the operator of the locomotive.

(8A) MAIN C.B. OPEN – RED
Indicates the main breaker has tripped for some reason.

- This indication is normal if no other indicating lights are on.
- If one or more other lights are on, refer to the corresponding description of effect and action.
  - When the "Main Circuit Breaker" is closed this light will go out.

(8B) OVERCURRENT BRAKE – RED
Indicates an overcurrent condition in the traction motor armatures during dynamic braking.

- Dynamic brake is disabled.
- Power contactors drop out.
  - Return throttle to "0" and depress FAULT RESET button.
  - Resume dynamic brake operation.
  - If fault repeats, dynamic brake must be cut out on the locomotive by moving the DYNAMIC BRAKE CUT-OUT switch to the ON position.

(8C) PCS OPEN – RED
Indicates that a penalty or emergency brake application has occurred.

- Propulsion system is disabled.
- Pacesetter is disabled.
- Power contactors drop out.
  - Return throttle to "0".
  - Recover air brakes by moving automatic brake lever to SUPPRESSION (penalty application) or EMERGENCY and hold for 45 seconds (emergency application), then move to RELEASE position.

(8D) PWR DEMAND LIMIT – WHITE
Power limit switch is in the ON position.

– Locomotive will be limited to one-half (1/2) power automatically.

– No action required.

(8E) APL NO PWR. – RED
Indicates any of the following:

1. High air temperature in the convertor.
2. Low oil pressure in the convertor.
3. Low oil pressure in the transformer.
4. High oil temperature in the transformer.
5. Phase fault in the convertor.
6. Auxiliary convertor is disabled.

– Auxiliary power convertor is disabled and all auxiliary motors shut down.
– If other indicating lights are on, refer to the effects and corrective action of the respective lights.

– Return throttle to the "0" position and depress FAULT RESET button.
– If fault repeats, isolate locomotive.

(8F) SAND – WHITE
Indicates that manual sanding is in operation.

– Sand is being applied in front of each truck.

– No action required.

(8G) WHEEL SLIP – WHITE
Intermittent flashing of WHEEL SLIP lights indicates moderate to severe wheel slip, in the back-up system of operation.

– This is an indication that the wheel slip control system is operating normally to control the slip.
– Buzzer will sound.

– No action is required unless severe lurching threatens to break the train; in that case throttle must be reduced slowly to regain control.
– Notify maintenance facility so that creep control failure may be corrected.

Continuous WHEEL SLIP light could indicate a locked wheel condition.

– A buzzer and alarm will be sounded.

– Refer to corrective action for "Locked Wheel" light.

(8H) LOCKED WHEEL – RED
Indicates a locked wheel condition, and/or a traction motor speed pickup failure.

– A buzzer and alarm will be sounded.
– WHEEL SLIP LIGHT will come on.

– Stop the locomotive immediately and make a thorough inspection to ensure there are no locked wheels.
– Railway procedures must be followed if a locked wheel is found.
– If locked wheel light is on due to a failure of the traction motor speed pickup, cut out the affected traction motor.

INDICATING LIGHT PANEL – LOCOMOTIVE CONTROL PANEL (Fig. 2-6)

(22A) GCB LOCKOUT – RED
Current in any of the main transformer secondary windings is excessive.

– Alarm is sounded.
– Main circuit breaker is opened.
– Main CB light is on.
– Pantograph is lowered.

- Power contactors drop out.
- Propulsion system is disabled.
  - Return throttle to "0" and depress the 86 RESET switch.
  - Depress ALARM SILENCER switch if necessary.
  - If fault repeats, isolate the locomotive.

Transformer oil pump circuit breaker is tripped.

- (same as above)
  - Reset circuit breaker.
  - (Same as above)

Fire extinguisher system is activated (20 second delay).

- (same as above)
- FIRE MACHINE RM. is on.
  - Investigate cause and severity of fire to determine if locomotive can resume operation.

(22B) PFC FAULT – RED
Indicates a ground fault is detected in any of the four power factor correction circuits.

- Affected contactor will drop out.
  - Return throttle to "0" and depress FAULT RESET button.
  - If fault repeats, use PFC switch to cut out affected circuit. Operation may be resumed but with reduced power factor correction.

(22C) GRD. RELAY – RED
An overcurrent and/or ground fault is detected in any of the traction motor armature circuits.

- Alarm is sounded.
- Main Circuit Breaker is opened.
- Power contactors are dropped out.
  - It will reset automatically within 10 seconds up to three times within a 20 minute time period.
  - After 3 reset attempts, automatic ground relay reset device locks out.
  - If the traction motor fault can be isolated, the GRD. RELAY RESET switch may be depressed and the FAULT RESET switch depressed to resume operation.

(22D) DB GRD RLY – RED
Indicates a ground fault in the dynamic brake grid circuits.

- No effect on operation.
  - Return throttle to "0" and depress the FAULT RESET button.
  - If fault repeats, move the DYNAMIC BRAKE cut-out switch to the ON position.

(22E) DB GRID BLWR – RED
Open circuit or short circuit in the grid blower circuit.

- Dynamic brake is disabled.
- Power contactors are dropped out.
  - Move the DYNAMIC BRAKE CUT-OUT switch to the ON position to cut out the dynamic brake.

(22F) AUX CONV FAULT – RED
Phase fault in the auxiliary convertor or the auxiliary convertor is disabled.

- Auxiliary convertor is disabled.
- Propulsion system is disabled.
- APL. NO PWR. light is ON.
- Power contactors are dropped out.

- Return throttle to "0" and depress FAULT RESET button.
- If fault repeats, a reset may be tried after a time that would allow the temperature to decrease sufficiently.
- If fault is cleared, operation may be resumed.

(22G) CONV HI TEMP – RED
High air temperature in the thyristor convertor.

- Auxiliary convertor is disabled.
- Propulsion system is disabled.
- APL. NO PWR. light is ON.
- Power contactors are dropped out.

  - Return throttle to "0" and depress FAULT RESET button.
  - If fault repeats, a reset may be tried after enough time to allow the air temperature to decrease.
  - If fault is cleared, operation may be resumed.

High oil temperature in the thyristor convertor.

- Armature current will be reduced automatically.

  - If light remains on, throttle should be reduced.

(22H) TM. BLR. STOP – RED
Either traction motor blower motor has stopped.

- Propulsion system is disabled.
- Auxiliary power is disabled.

  - Check Traction Motor Blower Circuit breaker and reset if tripped.

  NOTE: ADL must be running at full speed reset circuit breaker.

  - If circuit breaker cannot be reset, cut out 3 traction motors of the affected truck.

(22I) TRANS. THY. OIL. PR. – RED
Low oil pressure is detected in the transformer.
Low oil pressure is detected in the thyristor convertor.

- Auxiliary convertor is disabled.
- Propulsion system is disabled.
- APL. NO PWR. light is ON.
- Power contactors are dropped out.

  - Return throttle to "0" and if cause can be determined and corrected, depress FAULT RESET button and resume operations.

(22J) NO BATT. CHG. – RED
Battery charger is not operating.

- Alarm will be sounded.
- Batteries will begin to discharge.
- Normal operation may continue, until battery protection system is activated.

  - Depress ALARM SILENCER button if required.
  - Check for open battery charger circuit breaker.
  - If cause cannot be found operation can be continued until "Battery Protection" removes the batteries from control circuits battery voltage decreases below 55 volts.

(22K) MTS. LIMIT – RED
Traction motor thermal capacity is reached.

- Maximum current to traction motors will be reduced to the continuous rating.

  - No action required.

(22L) AIR COMP. – RED
The air compressor circuit breaker is tripped.

- Air compressor is shut down.
    - Reset circuit breaker.
    - If breaker trips again, operation may be resumed if operating in a consist, using the remaining compressors.

Hot oil temperature in air compressor.

- (same as above)
- Alarm is sounded.
    - Check oil cooler motor circuit breaker and reset if tripped.
    - Return throttle to "0" and depress FAULT RESET switch.
    - If fault is repeated, isolate the compressor.

Cold oil temperature in air compressor.

- (same as above)
    - Check if air compressor oil heater circuit is tripped.

(22M) MOTOR CUT-OUT – WHITE
One or more traction motors are cut out.

- Operation will continue at reduced power.
- Dynamic brake is disabled.
    - No action required.

(22N) GRD. RLY. FIELD – WHITE
A ground fault is detected in the traction motor field windings.

- No effect on operation.
    - Operation may be continued until ground can be repaired at the earliest time.

(22O) GRD. RLY. AUX. – WHITE
A ground fault is detected in the auxiliary power convertor circuit.

- No effect on operation.
    - Operation may be continued until the ground can be repaired at the earliest time.

(22P) BAT. PROT. TRIP. – WHITE
The battery voltage has decreased below 55 volts and the battery charger is not operating.

- Batteries are disconnected from the control circuits.
    - Check if the battery charger circuit breaker is tripped.
    - Set BATT. PROT. OVERRIDE switch to the ON position.
    - If battery charger starts supplying 74 volts DC as indicated by the battery charger ammeter, depress the BATT. PROT. RESET button and return BATT. PROT. OVERRIDE switch to the OFF position.
    - If the BATT. PROT. TRIP. light comes on again, the locomotive will have to be isolated.

(22Q) BAT. PROT. O'RIDE – WHITE
Indicates the BATTERY PROTECTION OVERRIDE switch is in the ON position.

- Battery voltage is reconnected to the low voltage circuits.
    - If the BAT. PROT. RESET switch, when depressed, can reset the battery protection system, the BATTERY PROTECTION OVERRIDE switch can be placed in the OFF position.

(22R) FIRE MACH. RM. – RED
Indicates a fire is detected in the machine room.

- Flashing red lights are ON in the machine room and at the access doors.
- A fire alarm is sounded.

- The pantograph is lowered.
- The MAIN CIRCUIT BREAKER is opened.
- The MAIN CB. OPEN light is on.
- After a time delay of fifteen (15) seconds halon gas is discharged in the machine room.

  - Check severity of fire and determine if operation may be resumed.

(22S) TR. TEMP. WARM – WHITE
Indicates a warm oil temperature in the main transformer.

- Armature current will automatically be reduced.

  - No correction required.

(22T) AIR COMP. HOT OIL – RED
Indicates an overtemperature of the compressor oil.

- Air compressor oil immersion heaters are turned off.

  - Check oil cooler motor circuit breaker and reset if tripped.

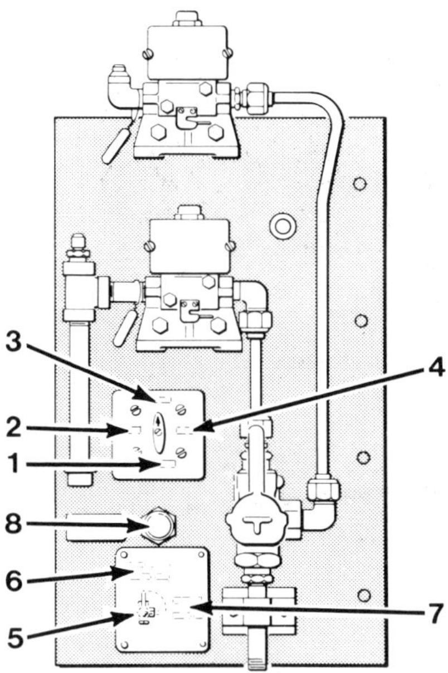

Fig. 2-1 Pantograph Panel

1. Electric Selector Switch
2. No. 2 Hood End
3. OFF
4. No. 1 Cab End
5. Pneumatic Selector Switch
6. No. 2 Hood End
7. No. 1 Cab End
8. Aux. Compressor Start

NOTE: IF BOTH SELECTORS ARE NOT IN THE SAME POSITION, THE PANTOGRAPH WILL NOT RISE WHEN THE SYSTEM IS ENERGIZED FROM THE CONTROL CAB.

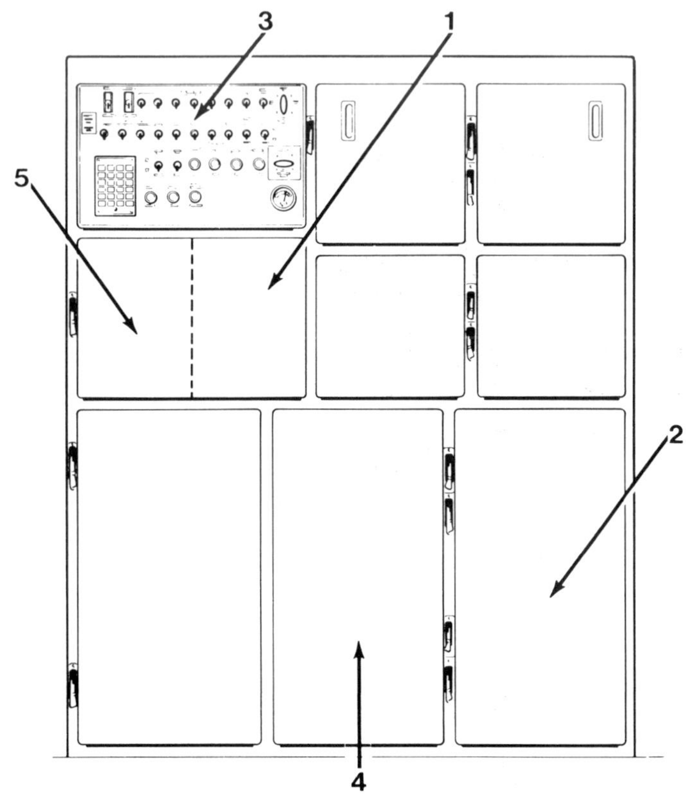

Fig. 2-3 Panel S7

1. 74 VDC Panel
2. 120/240 VAC Panel
3. Locomotive Control Panel
4. 480 VAC Panel
5. Switch and Fuse Panel

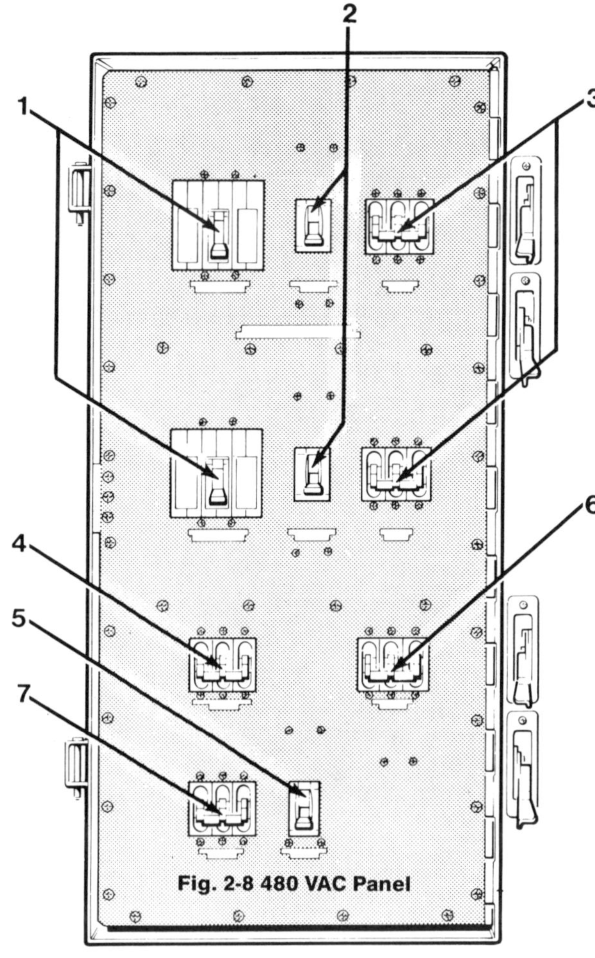

Fig. 2-8 480 VAC Panel

1. Two Thyristor Convertor Radiator Blowers
2. Two Traction Motor Blowers
3. Two Y1 Internal Blowers
4. Thyristor Convertor Oil Pump
5. Air Compressor Motor
6. Transformer Oil Pump
7. Air Compressor Oil Cooler

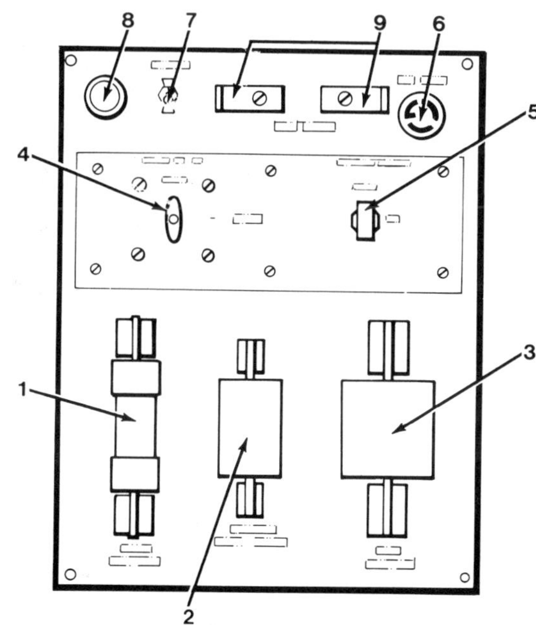

Fig. 2-9 Switch and Fuse Panel

1. Main Battery Fuse
2. Auxiliary Transformer Fuse
3. Auxiliary Locomotive Power
4. Ground Relay Cut-out Switcl
5. Battery Switch
6. 74 Volt Receptacle
7. Fuse test switch
8. Fuse test light
9. Fuse test blocks

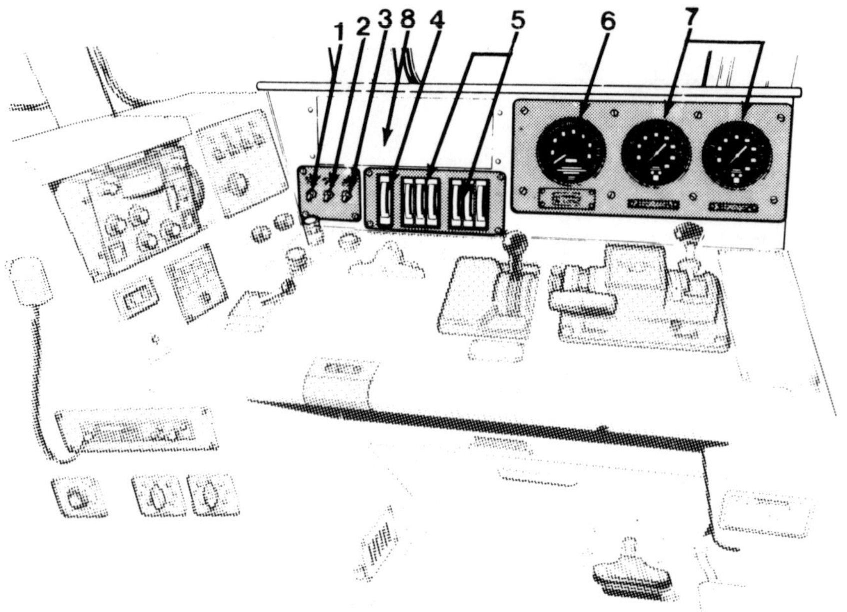

1. Gauge Lights
2. Reading Light
3. Ditch Lights
4. Dynamic Brake Ammeter
5. Traction Motor Ammeters
6. Speed Indicator
7. Air Gauges
8. LIC

Fig. 2-11 Front Vertical Panel

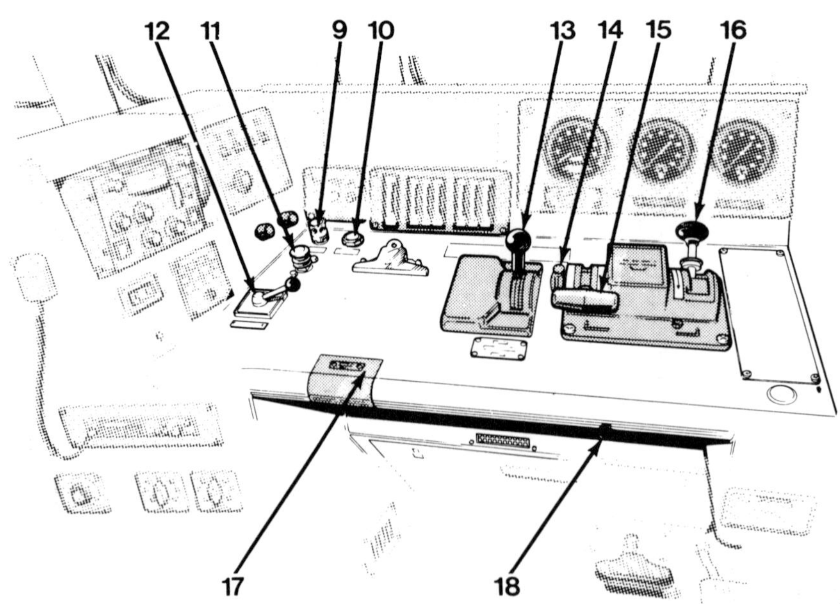

9. Alarm Silencer
10. Sanding Switch
11. Bell Switch
12. Reverser Control
13. Throttle
14. Cut-off Pilot Valve (auto brake)
15. Automatic Brake Valve
16. Independent Brake
17. Horn
18. Brake Pipe Pressure Regulator

Fig. 2-12 Front Horizontal Panel

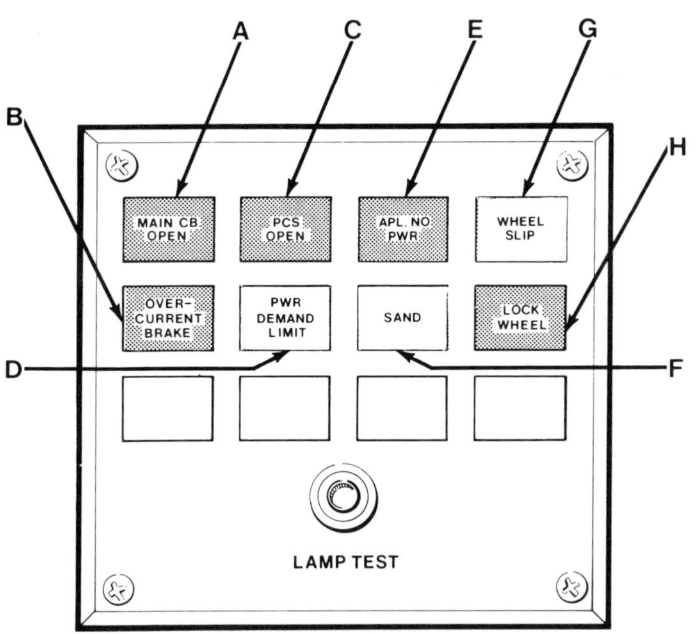

Fig. 2-17 Control Console (Fault Indicator Lights)

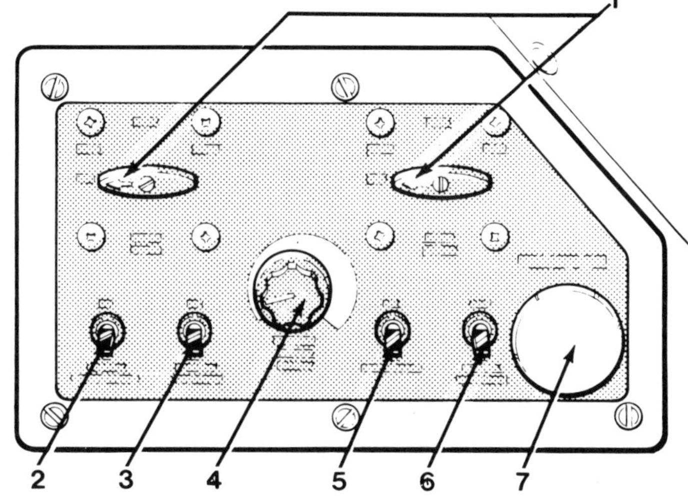

1. Front and Rear Headlights Selector Switch
2. Windshield Heater
3. Mirror Heaters
4. Gauge Light Dimmer
5. Pantograph
6. Circuit Breaker
7. Emergency Power OFF

Fig. 2-20 Switch Panel

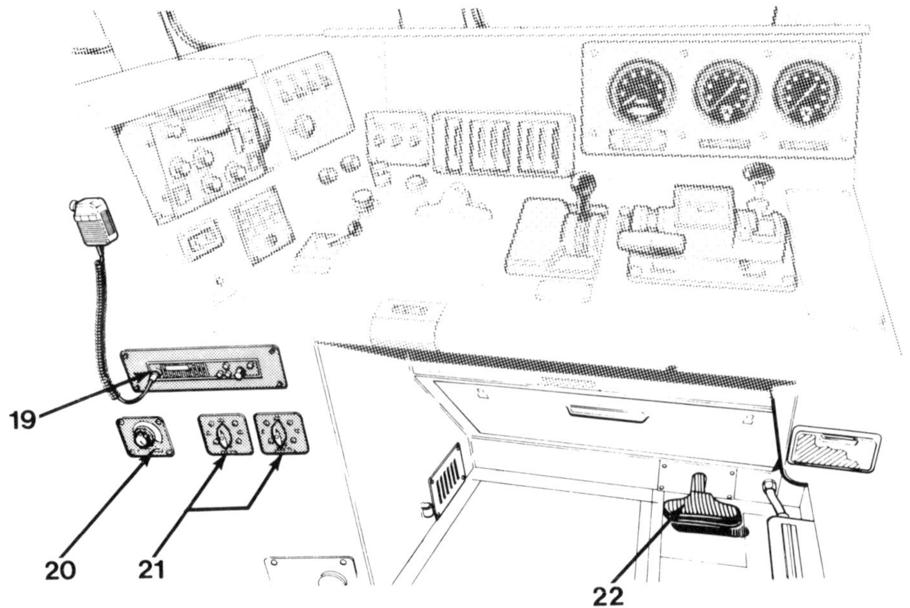

19. Two-Way Radio
20. Instrument Light Dimmer
21. Two Cab Heaters
22. Safety Control Foot Pedal

Fig. 2-18 Lower Section

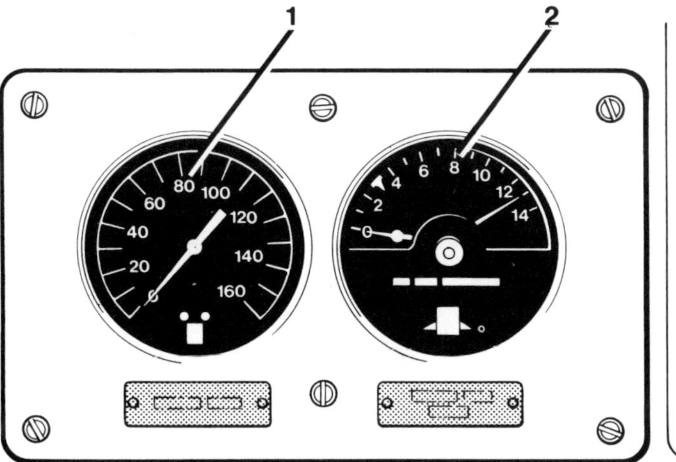

1. Air Flow Indicator
2. Signal Air Pressure Gauge

Fig. 2-22 Main Brake Panel

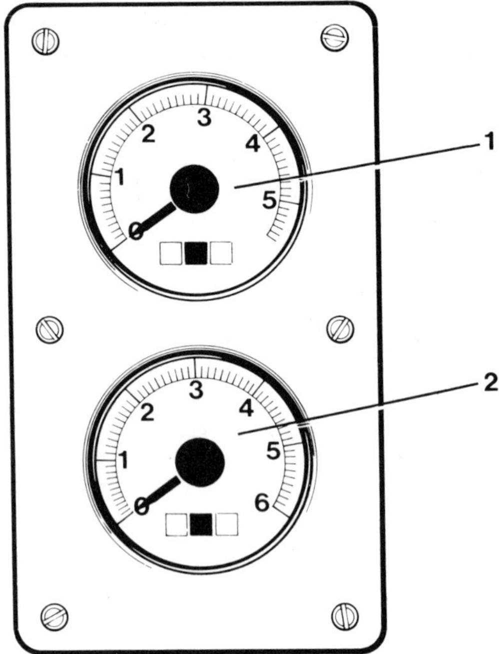

1. Power Indicator
2. Line Voltage

Fig 2-21 Electrical Panel

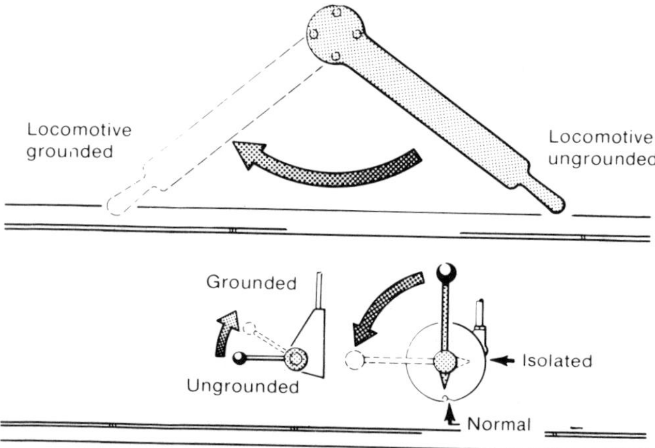

Fig. 3-1 Grounding Switch and Interlocks

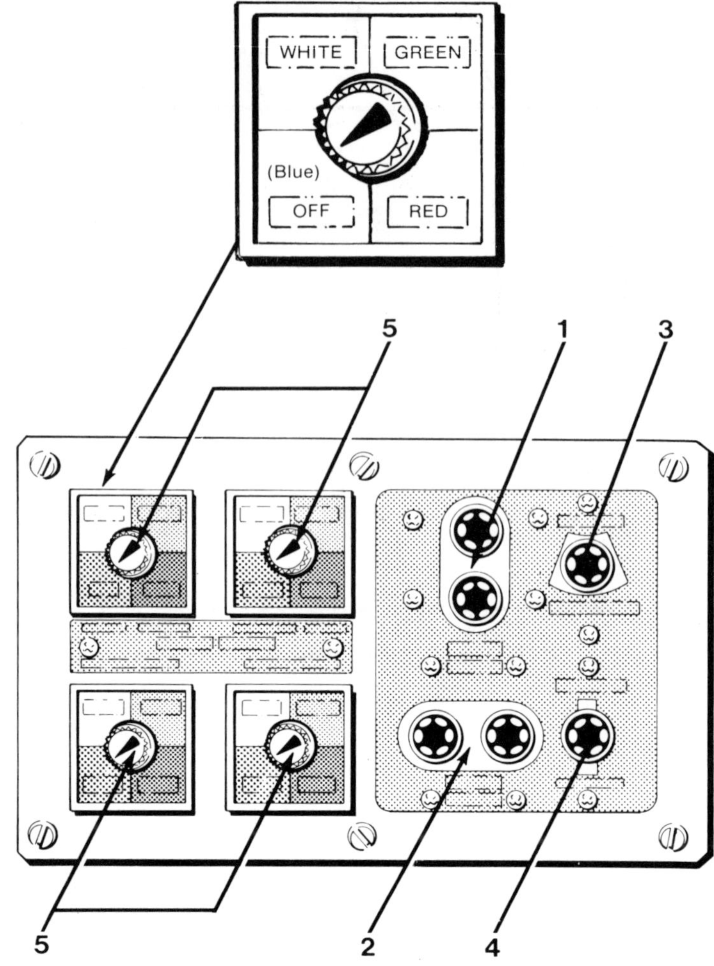

1. Two Front Headlights
2. Two Rear Headlights
3. Window Defrosting
4. Mirror Defrosting
5. Classification Lights Switches

Fig. 2-23 Classification and Indication Lights Panel

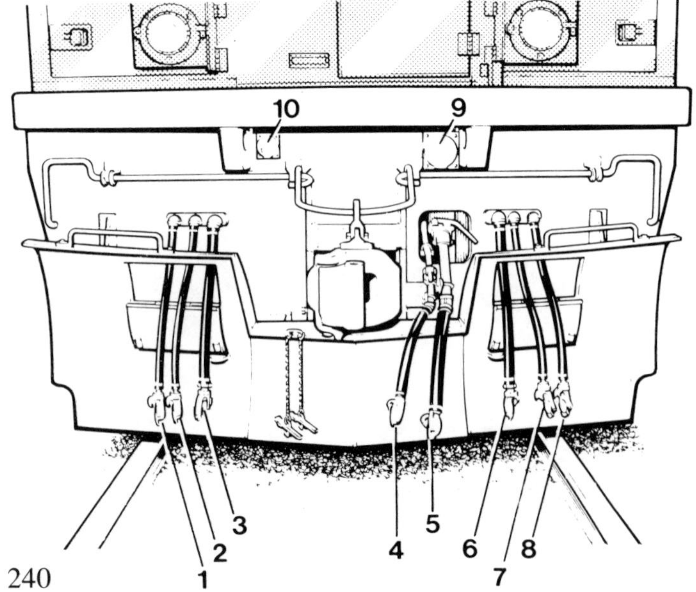

1. Independent Application/Release
2. Actuating
3. Main Reservoir Equalizing
4. Signal
5. Brake Pipe MU Receptacle
6. Main Reservoir Equalizing
7. Actuating
8. Independent Application/Release
9. MU
10. Snow Plow Lights

Fig. 3-2 Locomotive Hoses